Praise for Jona

'Jonathan Pinnock **writes compelling tales** with
a deliciously wicked glint in his eye.' Ian Skillicorn,
National Short Story Week

'Jonathan Pinnock is **Roald Dahl's natural successor**.'
Vanessa Gebbie

'**Funny, clever, and sometimes brilliantly daft.**
A comedy that I am sure would have made Pythagoras,
Archimedes and Douglas Adams all laugh out loud.'
Scott Pack on *The Truth About Archie and Pye*

Reader reviews of *The Truth About Archie and Pye*:

'Great fun. **A crazy roller coaster of a book**
that rattles along.'

'…**a wickedly clever story** that I thoroughly enjoyed.'

'Every page is amusing, **every chapter ends leaving you
wanting more**. Perfectly plotted.'

'Be warned, **the plot is bonkers**.'

'Who would have thought that **mathematics could
be so much fun**?'

'**Totally entertaining**… The plot is bizarre, strangely
plausible, and keeps the pages turning. And the
mathematical content is done in such a way as
to make you feel smug when you "get it".'

JONATHAN PINNOCK

BAD DAY
IN MINSK

A MATHEMATICAL MYSTERY,
BOOK FOUR

This edition published in 2021 by Farrago,
an imprint of Duckworth Books Ltd
1 Golden Court, Richmond TW9 1EU, United Kingdom

www.farragobooks.com

ISBN: 978-1-78842-303-8

To Gail, as always

'I really didn't want to go to Belarus.'

Tom Winscombe, *The Truth About Archie and Pye*

Author's Note

This is the fourth book in the *Mathematical Mystery* series that started with *The Truth About Archie and Pye*. When I was writing that first book, I decided that it would be cool to include a bunch of Eastern European mafiosi as part of the cast. I can't quite remember why I did this, although it may be related to my obsession at the time with the bizarrely overblown mausoleums favoured by Russian gangsters. Anyway, for some reason I decided that Belarus would be a suitable country of origin for these people, on the basis that I reckoned that most of my readers would have heard of it but very few of them would have visited it.

When it came to deciding what this fourth book was going to be about, I thought it might be interesting for Tom Winscombe to pay another visit to the Belarusian mafia, but this time on their home territory. Much of this book, then, is nominally set in Belarus, although I would be the first to admit that the Belarus depicted in these pages probably bears about as much resemblance to the actual one as Borat Sagdiyev's country of birth does to the real Kazakhstan.

What I hadn't anticipated was that at the precise time when I was due to hand in the manuscript to my lovely publisher, Belarus would suddenly begin to dominate the evening news, with a popular uprising calling for the end of the ghastly Alexander Lukashenko's repressive regime. It is therefore possible that there are those among you who have picked up this book anticipating a gripping political thriller set during the final days of the dictatorship, and I feel duty bound to inform you that it is in fact no such thing. It's just another daft comic caper that happens to be set somewhere in one of the countries that made up the former Soviet Union.

That notwithstanding, I would like to dedicate this book to the brave people of Belarus. Democracy is a fragile flower but one that deserves to be cherished.

Author's Other Note

As mentioned above, this is the fourth book in this sequence and a surprisingly large volume of adventurous water has flowed under the mathematical bridge since we first met Tom Winscombe. Being a reasonably conscientious sort of person, Tom does his best to paraphrase what has happened in the previous books where necessary for the benefit of new joiners. However, for readers who may require a little more in-depth background knowledge, I have begun to put together a Wiki called *Archiepyedia* for this very purpose. You can find it on my website at www.jonathanpinnock.com/wiki_archiepye/index.php/Main_Page. And don't worry, it has special buttons to protect you from spoilers!

Chapter 1

As instructed, Dorothy parked the car a couple of streets away. We were a minute or two early and there was a fine autumnal drizzle in the air outside, so neither of us was in any great hurry to get out.

'Are you sure about this?' I said.

'Why wouldn't I be?' said Dorothy.

'Because maybe it might be nice to relax and enjoy life for a little while?'

Dorothy looked at me and raised an eyebrow. 'Really?' she said. 'Really?'

I sighed. She was right, I suppose. We'd come too far to give up now. I checked the pavement for passing pedestrians and then opened the car door. Dorothy got out too and locked the doors.

'Come on,' she said, pulling up the collar of her jacket against the rain and marching off into the dusk.

The flat was situated over a dusty second-hand junk shop, and I couldn't resist looking in the window.

'Hey,' I said, 'we should come back in daylight and buy some of this stuff. Is that a Geiger counter or something there?' I was pretty certain it was, and you never know when you might need one of them.

But Dorothy was already pressing the buzzer at the side of the door. 'Focus, Tom,' she said. She was always the practical type.

A window opened above us and then slammed shut straight away before we had time to look up. The buzzer crackled and the door clicked open.

'Come on,' said Dorothy. I followed her up the stairs.

Katya showed us into the front room.

'Coffee?' she said.

We both demurred. When she had been our intern at Dot Chan, Katya had been an excellent developer but she was spectacularly bad at making coffee. Indeed, it wasn't until the previously healthy *Ficus elasticus* in the corner of the office died that it became apparent that everyone in the office apart from her had been emptying their mugs there. She herself had been completely unrepentant, claiming that she was merely being true to her Belarusian heritage by making coffee that way, although it was impressive that she still managed to do this with Nescafé Gold Blend.

Katya shrugged. 'My uncle will be with you soon,' she said and left the room.

The room was sparsely furnished, with a sofa and a couple of moth-eaten armchairs that exuded a vaguely musty smell, as if they'd been left out in the rain overnight a few months previously. The Anaglypta wallpaper had a thin sheen of nicotine to it and was torn in several places. An attempt had been made to cheer things up a little by putting up a poster advertising what I assumed was

'Minsk the Beautiful', although my Cyrillic was poor, so it might just as well have been advertising 'Minsk the Fairly Unpleasant' for all I knew. The carpet was dark with some kind of long-lost pattern that had blended over the years with a series of stains that could have come from food spillage, drinks or one of any number of bodily fluids.

I glanced at Dorothy, trying to give her an encouraging smile. She said nothing. After a few minutes, the door flew open and a massive, muscular figure entered the room.

'Arkady!' I said, standing up. He walked over to me and made a decent attempt to crush my hand into a fleshy sac stuffed with bone dust.

'Tom!' he said, slapping me on the back with his other hand. There was going to be quite a bruise there too. He turned towards Dorothy and feigned delight as if some goddess had dropped in. 'And your lovely woman Dorothy! She is still with you!'

'Don't sound so surprised,' I said.

'Ah, but you lose your women too easily,' said Arkady. I wasn't sure how to respond to this. My previous girlfriend, Lucy, had after all left me for him. But then again, he hadn't managed to hold onto her for very long either.

'You too, Arkady,' I said.

He grimaced and gave a sad shake of the head. 'Lucy was fine woman. Also very good at the sex.' He tilted his head on one side and looked me in the eye. 'Tell me, did she do that thing with you where she—'

I glanced nervously at Dorothy and held up my hand. 'I think maybe we need to talk about why you asked us to come here,' I said quickly.

'Yes,' said Dorothy. There was an edge to her voice.

'Yes,' said Arkady. 'Yes, of course. I am sorry. But I miss Lucy. She happy? I hear she engaged or something.'

'She's with some junior proctologist,' I said. I was about to go on to describe how I ended up accidentally impersonating him when I joined a stag party on the artificial island of Channellia just before I unwittingly set it circulating on a course based on the Fibonacci sequence, leading to it collapsing into the sea five miles off the coast of Burnham-on-Sea. But then Dorothy gave a discreet cough and I stopped myself just in time.

'Arkady,' she said. 'Katya said you mentioned something about the Institute for Progress and Development.'

'Ah, yes. I am sorry,' he said. 'I am leading difficult life at the moment. I do not see many people I know and when I do, I ask too many questions.' He looked sad and lost for a moment, and it struck me that, underneath all the bluster, Arkady didn't seem quite the man he once was.

'Where should I start, then?' he said.

'Maybe at the beginning?' said Dorothy.

'Ah, if I knew when that was, perhaps. But no.'

'Maybe you can tell us what happened to the Vavasor papers, and then go on from there.'

I sighed. It was always going to come back to the Vavasors, wasn't it? The mathematical papers belonging to the long-dead twins Archie and Pye Vavasor had attained something of a mythical status for Dorothy. I'd briefly ended up with them in my possession after bumping into the Vavasors' biographer on the train, shortly before he was murdered by the rogue financier Rufus Fairbanks. Then Dorothy was kidnapped by the

Gretzky gang, who Rufus Fairbanks had previously assisted with money laundering using algorithms developed by the Vavasors – at least until both Vavasor twins had independently had affairs with Fairbanks's wife Cressida, leading to Archie killing Pye and then subsequently killing himself. So I ended up trading what I thought were the papers for Dorothy, although it turned out that the briefcase containing them had also been packed with high explosives by Arkady's friend Sergei, who had something of a grudge against the Gretzkys because they had killed his brother Maxim.

'Hold on,' I said, coming out of my reverie. 'Surely the papers were blown up in the explosion?'

'Tom,' said Dorothy, 'we've been through this. Sergei could have removed them first.'

'He did,' said Arkady. 'Remember, he work for Isaac at the time.' Oh god. I'd forgotten that bit. Isaac Vavasor was the twins' younger brother. He was now dead himself, killed by one of the Fractal Monks while trying to escape from their monastery in Greece.

'So, hold on, then. Does that mean that the papers simply went back to Isaac?'

'Not quite. Sergei didn't get the papers back to Isaac. He disappear as soon as he hear you successfully deliver bomb to Gretzky gang.'

'Can I just point out that I had no idea I was delivering a bomb?' I said. I didn't like the idea of being some kind of terrorist. It really wasn't my style.

'Sssh, Tom,' said Dorothy. 'Carry on.'

'I disappear too, of course,' said Arkady. 'Is risky for me too.'

'That's something I've never really understood,' I said. 'How come no one associated with the Gretzkys came after us?'

'No one knew you involved. Gretzkys are small operation here and not many left after—' here, he mimed an explosion. 'The only car anyone saw driving away was the one you borrow from me, Tom.'

'But they expected trouble from Sergei.'

'Yes. And they know of me, too. It is small community.'

'So what happened to the papers?' said Dorothy, with more than a hint of desperation in her voice.

'Ah,' said Arkady. 'As I say, Sergei disappear. I not know where he is and I not hear anything for week or two. Then he call me. He say he worried he being followed.'

'And I guess he would know,' I said. From what I knew of Sergei, his primary skill set was counter-intelligence.

'He would, Tom. He would. So next day I get another call and he tell me he find out who following him. He – what's the word? – he turn the table, he follows the follower. And guess where he come from?'

Dorothy and I looked at each other, then shrugged.

'He come from this Institute for Progress and Development,' said Arkady, triumphantly.

'So not the Gretzkys?' said Dorothy.

'No, not the Gretzkys.'

'But why the Institute?'

The Institute for Progress and Development was one of those shady libertarian think tanks with a sketchy funding stream and an inordinate amount of influence on government policy. Although in their case, Dorothy and I did at least know where some of their money came from.

'Are they still around?' I said. 'I thought they went down with Channellia.'

'The Institute's still there,' said Dorothy. 'Only last week they brought out a paper advocating zero taxes on tobacco to boost the economy. Got a lot of publicity.'

'Oh, that was them, was it?' I remembered wondering who was managing their PR. However you felt about the morality of it, it was impressive stuff. 'But Channellia was their main source of income, wasn't it? The cryptocurrency scams and drugs were all run out of there, and now it's sitting at the bottom of the Bristol Channel.'

'Well, they're still getting money from somewhere,' said Dorothy. 'But that's neither here nor there,' she added. 'Why were they following Sergei?'

'Ah, well this is where it get difficult,' said Arkady. 'Next day, Sergei calls me again, say he has new job and he have to leave country. Won't say where he go.'

'So what happened to the papers?' said Dorothy.

'Last week, I get call from this lady Carla. She say she Sergei girlfriend and she very worried. She come home three days ago and flat all a mess.'

'Oh no,' said Dorothy.

'And here's interesting thing. Before Sergei go away, he show Carla important papers he keeps under mattress. Sergei say if anything strange happen, burn them. So she go to find papers and they gone.'

'Gone?'

'Gone.'

'The Institute must have stolen them,' said Dorothy.

'I think you right.'

'But why would they want to steal a bunch of obscure mathematical stuff?' I said.

'The same reason as everyone else,' said Dorothy. 'There are some valuable algorithms in those papers. Probably.'

'Probably?'

'Well, I didn't get much chance to look at them properly, did I?'

The fact that the extreme complexity of the mathematics involved, combined with the Vavasors' appalling writing and idiosyncratic presentation had baffled even Dorothy still rankled with her.

'So are we saying that the Vavasor papers could be in the hands of the Institute for Progress and Development?' I said.

'I think yes,' said Arkady.

'And we still don't know where Sergei is?' said Dorothy.

'No. Carla not know either.'

'She doesn't know?' said Dorothy.

'No. Sergei very secretive guy. He good guy but he keep everything close. Too close.'

'But what are we going to do?' I said. 'Break into the Institute and steal the papers back?'

Everyone looked at me in silence.

'I think that would be an excellent idea,' said Dorothy eventually.

Chapter 2

I stared at Dorothy. 'Really?' I said. 'First of all, we don't even know where the Institute for Progress and Development is. Remember? All we managed to find last time we looked was a forwarding address.'

Arkady held up his hand. 'Is OK. Sergei tell me where it is. Is in Westminster. Near the House of Parliament. I have address.' He began to stand up.

'No, it's OK, Arkady,' said Dorothy. 'We don't need it quite yet.'

'No we bloody don't,' I said. 'Can I continue? Secondly, we don't know for sure that the papers are actually at the Institute. Thirdly, we don't even know if there's anything worthwhile in them anyway. And finally, and I know this is a really, REALLY minor point, but last time I looked, burglary was illegal.'

'Yes, but—' began Dorothy.

'Oh, and one more thing,' I said. 'If we were to get caught, you know what else would happen? They'd take our fingerprints. And guess what else might turn up then?'

Dorothy sagged slightly and was quiet for a moment. Our fingerprints were all over Rufus Fairbanks's kitchen on the night that he was killed in a bizarre food mixer-related incident. The last thing we wanted was to be put in the frame for that one, even if we weren't technically responsible. I didn't fancy trying to explain that away under cross-examination.

'We won't get caught,' she said eventually.

'And that's it? That's your solution? We won't get caught? Come on, Dorothy.'

'We need those papers.'

'You need them,' I said. 'Not me.'

'Come on, Tom, you've come as far down this road as I have.'

I shook my head. I'd had enough of all this. 'How are we going to do it anyway?' I said. 'Won't they have security systems and stuff?'

'Of course,' said Dorothy. 'But we can hack them. *I* can hack them. Remember, I'm good at that.'

'Yeah, but what if they use plain old-fashioned locks?'

'We'll deal with that if and when it arises.'

'No, I'm sorry. It's time for all this to stop. It's all too risky.'

'I'll have to do it on my own, then.'

A chilly silence filled the room. Then Arkady spoke.

'I will help you,' he said.

'Thank you,' said Dorothy.

'What?' I said. I glared at Arkady. No way was this overdeveloped muscle bag going to take another of my girlfriends away from me.

'Is OK,' he said, 'I not take your Dorothy. She not my type.' I felt myself reddening. Was I really that transparent? I also felt vaguely insulted that he didn't fancy Dorothy.

'I'll help you too,' came another voice. Katya padded into the room and perched on the edge of the sofa next to Arkady. She had evidently heard every word of our conversation.

'What?' I said again. I could feel myself being outmanoeuvred in real time.

'It's OK, Tom,' said Dorothy with a teasing smile. 'Three should be enough.'

'No,' I said. 'We can't involve the kids. I mean, sorry Katya, but as your employer, Dorothy has a duty of care to you. I'll go instead.'

'That's not fair,' said Katya. 'And in any case, I'm back at uni now. So I can do as I please.'

'Well, just don't get Balvinder involved as well,' I said. Balvinder was the other intern.

'I've already texted him,' said Katya. 'He's up for it.'

'Oh, come on,' I said.

'You don't know Bal very well, do you? His party trick is picking locks.'

'Great,' said Dorothy. 'So when are we all going in?'

Three o'clock in the morning, two nights later, the five of us were lurking in an alleyway opposite the Regency town house that constituted the headquarters of the Institute for Progress and Development. At least, we were lurking opposite the address that Arkady had been given by Sergei, even if there was nothing resembling a name plaque to indicate what went on inside the building. The place was in darkness and there was no evidence of any nocturnal activity.

'Nice place,' I said. 'Wonder what the rent is?'

'They probably own it outright,' said Dorothy. 'Or at least one of their nought point nought nought one percenter friends does.'

'Are you sure you got the right number of noughts there?'

'Maybe a couple more. Who knows?'

'So what's the plan?'

Dorothy pulled out her phone and raised her hand.

'Give me a moment while I get into the Wi-Fi.' She began tapping away, muttering to herself as she did so.

After a while, I began to get bored.

'So how did you end up picking locks as a hobby?' I said to Balvinder.

He shrugged. 'Dunno, really. Just sort of happened. Was messing about with an old padlock and found I could open it without the key.'

'Ever thought of taking it up as a career?'

'What?'

'You know. Safe-breaking. Burglary.'

'Who says I haven't?'

I looked at him. He was completely expressionless. Then he burst out laughing. 'Oh, mate,' he said. 'Your face. You should just see your face. How do you think the Aunties would feel if they found out their precious Balvinder had taken up a life of crime? Believe me, it's bad enough that I'm doing a PhD.'

'At least you'll end up as a doctor,' said Katya.

'Not a proper one in their eyes.'

We were interrupted by a suppressed howl of rage from Dorothy.

'You OK?' I said.

'Yes, Tom,' said Dorothy. 'Now shut up, please.'

Katya wandered over and began peering over Dorothy's shoulder, who turned round and glared at her. Katya put her hands up and moved away again. Dorothy carried on tapping and then finally raised her fist in the air. 'OK,' she announced. 'I'm in. No 2FA, thank god.'

'2FA?' I said.

'Two factor authentication,' said Katya. 'Sends a message to one of their phones to confirm we're valid. Which would have screwed everything up.'

'Great,' I said. 'So can we switch all the cameras and everything off?'

'Almost there,' said Dorothy. 'Just need to upload this little widget that'll keep replaying the last hour of video while we're in there.' She tapped a couple more times and then said, 'OK, it's done. Let's get moving.'

We all crossed the road, Dorothy leading the way and Arkady bringing up the rear. When we were all safely over, Arkady stepped back into the road and checked both ways.

'All clear,' he said.

'OK, guys,' said Dorothy. 'Which way in, do you think? Front door or basement?'

'Gotta be basement, surely,' I said.

'Basement it is, then,' said Dorothy. 'Come on, Bal. Do your stuff.' She stepped out of the way to let Balvinder go down towards the basement entrance. Arkady and Katya remained at street level to keep a lookout.

Balvinder took his rucksack off, knelt down and examined the lock. After a while, he gave a little chuckle and stood up.

'Dear oh dear,' he said. 'Basic BS3621 five lever mortice deadlock.'

'Is that good or bad?' I said.

'Oh, it's good. Shouldn't take more than a couple of minutes.' He opened up his rucksack and fished inside, eventually coming out with a small unbranded black box.

'What the hell's that?' I said.

'Thing I invented myself. Laser. Aim it into the lock so it bounces around and comes back with a reading of the structure. Connected to my phone over Bluetooth.'

'Neat,' said Dorothy. 'So you don't use stethoscopes and stuff any more?'

'Nah. That's old school.'

What was it about mathematicians and computer types? I remember the scandal in my final year at uni, when several of the exam papers went missing from the central office on the day before they were due to be sat. No one was in any doubt as to where they needed to look to find the culprit, and it certainly wasn't in the Arts school.

Dorothy used to claim it was because they spent so much of their lives trying to solve problems that it affected their entire outlook on life. Everything became something to hack. My view, on the other hand, was that most of them were just high-functioning sociopaths who felt they were above any of the rules that applied to normal people. As with so many things, Dorothy and I failed to find common ground here.

'OK, here goes,' said Balvinder. He held his box up to the lock and got out his phone. He located an app, pressed it and a display appeared showing a percentage spinning upwards towards 100 while a circle was drawing itself around the outside. When it got to 100 per cent and the circle was complete, the phone beeped softly a couple of times and the display changed to show the outline of a key, with numbers assigned to each part.

Balvinder removed the box from the door and put it back in his bag. Then he took out what looked like a key, except that each of the five pins seemed to be adjustable. He looked at the results from the scan and started moving the pins around, fixing them in place with tiny grub screws. When he was happy, he took one last look at the key he'd created and then turned to us.

'OK, what do you reckon?' he said.

'Looks good to me,' I said.

'Me too,' said Dorothy.

'Let's give it a go, then,' said Balvinder. 'Give the lock a quick clean first, though. You'd be amazed at the rubbish that collects in these things.' He reached into his bag and took out an airgun and a can of WD40. He squirted the lock with a blast of air and then gave it a good squirt of lubricant. Then he put them back in his bag, took the key and gently inserted it.

'Dorothy?' he said, motioning towards the lock. 'Fancy doing the honours?'

'You just want me to be responsible for breaking in,' she said.

'Well, yeah, there is that. But I thought you'd like to be first in, too.'

'Cool.' She took hold of the key and began to turn it. It seemed to meet no resistance at all and a moment later the door was unlocked. She turned to us and gave a broad smile. 'Good work,' she said to Balvinder, who gave a shy nod.

'You definitely sure you've switched everything off?' I said.

'Have I ever let you down?' said Dorothy.

She pushed open the door. We were in. We called to Arkady and Katya to join us and we all filed inside the basement, switching our torches on once we were inside.

I wasn't sure what we were supposed to be expecting, but on reflection I guess I'd been anticipating a bit more order about the place. Or maybe that was the libertarian way. As our torches scanned the cavernous room, they lit up teetering piles of books, bulging lever arch files and random bits of old computers, including several printers, some of which looked as if they'd been subjected to a right good kicking. Over to one side of the room, there was a detailed scale model of Channellia, which I guess could come in handy should they ever decided to resurrect the concept.

Also, right next to where I was currently standing, Dorothy's torch suddenly caught a large stuffed gorilla, which I definitely hadn't been expecting.

'Jesus!' I exclaimed, stepping sharply back, arms flailing. As I moved, my right hand caught Arkady, knocking the torch out of his hand, straight towards a large and exceptionally ugly ornamental vase. There was a loud crack, then a wobble, followed by an excruciatingly loud crash as it hit the floor.

All the torches went out and there was complete silence for thirty seconds, overlaid only by the sound of five people not breathing.

'Tom, you idiot,' hissed Dorothy, eventually. 'What in god's name did you do that for?'

'What?' I said. 'You shone the torch at the gorilla. And what's a fucking gorilla doing here anyway? Or a stupid vase?'

'Why shouldn't there be one?'

'Because normal people don't leave stuffed gorillas lying around in their basements. And in any case, was I the one who was waving the torch around? And speaking of torches, I don't think it was mine that broke the vase, was it, Arkady?'

'Tom,' said Arkady, in a voice that was much more reasonable than it should have been. I decided it wouldn't be useful to my cause to bring him into it.

'You know, I wish we hadn't brought you along, Tom,' said Dorothy.

'What?' I said. 'But I didn't even want to come along! I wanted a quiet night in. Netflix and stuff.'

'Yeah, but you tagged along anyway, like a bad smell.'

'Hey, that's unfair. You didn't give me the option.'

'Look,' said Katya, 'can you two stop it? It's like having your parents arguing.'

'Yeah,' said Balvinder. 'I feel like I want to go and hide in my room.'

'Also, we need to keep noise down,' said Arkady, casting his eyes towards the ceiling.

We all stopped and listened.

'Did you hear that?' whispered Dorothy.

'What?' I said.

'I hear nothing,' said Arkady.

'No, there,' said Dorothy. 'Again.'

We all strained our ears once more. I still couldn't hear anything.

'Something moved up there,' she said. 'I'm sure of it.'

'Maybe they'll think it was rats or something,' I said. 'Outsize rats.'

Even in the dark, I could sense Dorothy rolling her eyes at me.

'I could squeak a bit,' I said. 'Just to complete the impression.'

'Shut up, Tom,' she said.

We all waited in darkness for another minute. There was no noise from above us and even Dorothy agreed.

'OK,' said Arkady. 'No one coming I think. We lucky.'

There was a sound of exhaling from Katya and Balvinder. We all switched our torches back on again.

'But what we going to do?' said Arkady. 'This place is mess.'

'It's been pre-ransacked,' said Dorothy.

'Yeah,' said Arkady. 'So where we start to look?'

'Not worth even starting,' said Dorothy. 'There's nothing to see down here.'

'How can you be sure?' I said.

'Look around you,' she said, waving her torch about the place in a sweeping arc, lighting up, among other things, two pinball machines, a manual lawnmower and an ornate four-storey doll's house. 'Everything is covered in several months' worth of dust. Unless someone's gone to great trouble to replace it – and I really don't think that's their style – nothing new's been dumped here for several weeks.'

'Fair point,' I said. 'Upstairs, then?'

'Guess so,' she said, with an anxious upwards glance.

There was a staircase in the middle of the basement that went up to a closed door. I could see a thin, watery ribbon of light seeping in under it, presumably from the moon shining into the hallway above us. Dorothy led the way and I heard her muffle a curse as she tried the handle.

'Shit,' she whispered. 'It's locked.'

There was some muttering behind me and then Balvinder pushed his way to the front.

'OK,' he said, 'let me take a look.'

Dorothy stepped to one side to let him through and he knelt down to examine the lock.

'Ha,' he said after a while, rubbing his hands. 'Key's still in the lock. This is proper old school stuff.'

He bent down and rummaged around in his rucksack, taking out an old copy of a magazine. He eased this into the gap under the door and positioned it beneath the lock. Then he took out a wooden-handled tool like an awl and began fiddling around in the lock. Eventually there was a soft thud on the other side of the door.

'Now for the moment of truth,' said Balvinder, gently pulling the magazine back towards him until it came free.

'Isn't the key supposed to be sitting on it?' said Dorothy.

Balvinder held up a hand. 'Don't panic. Just a bit too thick.' Then he took his awl and wiggled it around under the door until part of the key was visible on our side. He grabbed hold of it and wiggled it back and forth until it finally came clear.

'There you see,' he said, with an air of triumph. This time, he wasted no time at all in putting the key in the lock and opening the door. I think he sensed that we were all feeling a little tense by now.

Now that the door was open, we all bundled after him into the hallway and stood for a moment, adjusting our eyes to the dim light and listening for any sounds. The hairs on the back of my neck were on high alert, although for one thing, I was fairly sure that they were simply reacting to my overactive imagination. More importantly, I frankly have no idea what they intended to do in the event of any kind of confrontation. I have always suspected that in the end, neck hairs are guaranteed to let you down in a fight and should not be relied upon for support.

'OK,' I whispered. 'What now?'

'We search the place,' said Dorothy.

I tugged at the door handle opposite us. 'Going to take a while if we're going to have to keep waiting for Balvinder to get us in to each room.'

Dorothy went silent.

'Then again,' said Balvinder, emerging from a cupboard further down the hall, waving a large metal ring with several sets of keys dangling from it, 'we could use these instead.'

'Nice work,' said Dorothy.

'Even better,' said Balvinder, 'they're all grouped into different floors.'

'Brilliant,' said Dorothy. 'So let's split up.'

'Jesus,' I said. 'Did you really say that? It never ends well when they do that in the movies.'

Someone – possibly Katya – suppressed a snigger.

'This isn't the movies, Tom,' hissed Dorothy.

'Same principles apply,' I said.

'Well, what else do you suggest? We need to search the whole building and we need to do it fast so we can get out before we get caught. If anything happens, we communicate it to the others via WhatsApp.' We'd set up a temporary group for the operation, which I'd named 'Raiders of the Lost Institute.' None of the others found this the slightest bit funny.

'Seems good plan,' said Arkady. 'I take ground floor.'

'OK, you and Balvinder take the first floor,' said Dorothy, nodding at Katya. 'Tom, you take the top floor, and I'll check out the area at the back.' Balvinder handed me a set of keys.

'Cool,' said Katya. Then she paused, before adding, 'What are we looking for exactly?'

'Bunch of papers with a load of mathematical scribbles on them and a bunch of cryptic comments in the margins,' I said. 'Plus a big glühwein stain on the front page,' I added. 'Long story.'

The glühwein stain was part of the mythology of the Vavasor papers, having been unwittingly deposited there by the careless Dolmetsch brothers on one of the rare occasions when outsiders had been permitted a glimpse. I always felt an affinity with the Dolmetsches. I wasn't safe with a full wine glass either.

'Okaaaaaay,' said Katya, not sounding entirely convinced.

'You can't miss them,' I reassured her.

'Well, with any luck someone else will find them,' she said, breezily.

'Or no one at all,' I said.

'Cheers,' said Dorothy. 'And Tom,' she added, 'try not to break anything, OK?' Then she headed off down the hallway towards the back of the building. Arkady slipped his way into the room at the front, and I followed Katya and Balvinder up the stairs.

After we parted company, I continued on up and those neck hairs started jangling again. Come on, Tom. This was getting silly. I paused at the top of the second flight of stairs and strained my ears to hear any movement. But all I could hear was the muffled noise of Katya and Balvinder moving around one floor below me. I shook my head to clear my thoughts and tried to orient myself. This was going to be completely straightforward. Just check all the rooms up here, find the papers and get out.

No problem at all.

Chapter 3

There was just a single door in the wall at the front of the building, with a portentous wooden plaque in the middle of it bearing the legend 'Boardroom.' At the back, facing me now, were three doors to choose from, and then finally, a further three doors in the walls on each side of me. Every single one of them was closed. I decided I would check each one in turn, clockwise, finishing up at the front.

The first one was unlocked and opened up into a tiny windowless kitchenette. I switched on my torch and scanned the room. On the work surface, there was a microwave and one of those coffee machines that require you to take out a subscription to a manufacturer of unrecyclable plastic pods. This would have been just the Institute's style: unnecessarily expensive and gratuitously bad for the environment. There was also a tin full of luxury biscuits, which I decided to sample, partly for research reasons but mainly because I was feeling a bit peckish. Also, I was learning that when you got involved in this kind of marginally dangerous operation, you could never be quite sure when you were going to get fed next, and breakfast was still a long way away.

The biscuit was a quite exceptionally pleasant chocolatey number, some kind of high-end synthesis between a Bath Oliver and a Hobnob, so good in fact that I had one more, just to be completely sure. The second one confirmed my positive opinion of the initial offering. These people might have dubious politics, but they appeared to have some taste in at least one key area.

I was now wondering if they had any cake, because the quality of the contents of the biscuit tin suggested to me that if there was indeed any cake on offer, it was going to be of excellent quality. However, after searching through all the cupboards and finding out that the only thing they contained was crockery, I realised I was bound to be disappointed. Maybe they just bought in a cake whenever they needed one in the boardroom and then scoffed the lot without leaving any. It's what I'd do, after all.

I was getting distracted.

It was touch and go, but after an appropriately Nietzschean effort of will, I eventually forced myself to leave the kitchenette without even so much as a glance at the contents of the fridge and moved on to the room next door. I found the right key after a couple of attempts and found myself in a small office with a couple of desks facing each other, illuminated by the moonlight shining in through the window opposite the door. Both desks were covered in a finely crafted mess of expense claims, restaurant receipts and illegible lists of Things To Do. There was a whiteboard on the wall, on which was written some kind of important exhortation, which I couldn't read because the letters were Cyrillic.

There was a filing cabinet in the far corner and I suddenly remembered what it was that I was supposed to be doing here. I walked over to it and gave the top drawer a good tug. Frustratingly, it was locked shut. I fished out the key ring from my pocket and went through the ones I had at my disposal. None of them seemed to fit a filing cabinet, however. I would have to go and get Balvinder to deal with it.

No, wait.

Sod Balvinder. I didn't need him and his fancy tools. Anyone could break into a filing cabinet. I'd seen it done in some crime thriller or other. You just got a ruler or something and slid it along the top of the top drawer until you triggered the catch.

Easy.

I fished around in the drawer of the desk near me and managed to find a ruler. I inserted it between the lip of the uppermost drawer and the casing and slowly moved it along. Halfway along, it met some resistance. Right, this was it. I pushed away at where I thought the catch must be and absolutely nothing happened. I pushed harder. Still nothing. I pushed harder still and this time, the ruler broke in two and half of it pinged off towards the ceiling, leaving the rest inside the drawer.

Shit.

I hunted around again and this time I found a metal ruler in the drawer of the desk opposite me. I was certain this would do the trick, and sure enough it did. The problem was, however, that the top drawer sprung out at high speed, toppling the cabinet towards me. The second drawer down decided to join in the fun as well, accelerating the cabinet's trajectory towards the ground. I staggered back towards the chair on the far side of the desk, intending to lean on it for security.

Unfortunately, the chair had other, more playful, ideas and spun backwards towards the opposite corner, taking me and the tumbling cabinet with it. Eventually it left us behind completely, and I ended up crashing towards the floor, with the filing cabinet clattering on top of me, spilling its contents in all directions as it did so.

I lay there for a moment, pinned to the floor and wondering how to extricate myself without adding to the pandemonium. Then I felt something vibrate in my trouser pocket. With considerable effort, I managed to extricate my phone, only to see a WhatsApp message from Dorothy.

Was that you, Tom? she wrote.

What? I replied.

That bloody great crash.

I was two floors up from her, so provided I managed to get everything back in order before we met up again, there was no way she could verify whether or not I was telling the truth.

I have no idea what you're talking about, I said.

You so do, Tom. Idiot.

This seemed unfair to me. *Innocent until proved guilty, eh?* I replied.

Wasn't us, said Katya, unhelpfully.

Cheers, I said.

Is not me, said Arkady.

Could be rats, I said.

Didn't hear any squeaking, said Dorothy. *Don't try to get out of it, Tom.*

How did she know it was me? It could easily have been Katya or Balvinder. I thought of pointing that out to her, but when she was in this kind of mood, it was pointless. If anything bad happened, it was always bound to be my fault. I mean, obviously this time it actually was, but that was beside the point. There was a principle at stake here.

I stuffed the phone back in my pocket and crawled out from under the filing cabinet. I stood up and checked myself for any damage. Nothing seemed to be broken, although there were a few cuts and bruises on my hands in particular. I righted the filing cabinet and located it back in its correct position in the corner. Then I knelt down again and scanned the floor with my torch. There were papers scattered everywhere and there was absolutely no chance of me managing to put them back in the right order.

On reflection, there was a decent chance that no one looked at them much anyway, so I just stuffed them back into the suspension files as best I could. Needless to say, there weren't any mathematical papers in there at all. They were mainly a load of what appeared to be contracts, some in English, some in some Eastern European language, along with bills for large quantities of building materials and equipment.

Once I'd packed the last of the papers away, I gave the top drawer a good slam and the filing cabinet seemed to lock itself again. I checked the state of everything else in the room, and when I was satisfied that I was leaving it in something close to the state in which I found it, I exited the room and pulled the door closed behind me, turning the key in the lock as I did so. Only eight more rooms to check.

The next one was easier to deal with. The only furniture in the room consisted of a couple of chairs along with a camera on a tripod pointing straight at them. Behind the chairs was a bookcase filled with political biographies, contrarian tracts and the complete works of Ayn Rand. I guessed this was where they prepared their YouTube videos and also where they broadcast their pundit spots on late night news programmes.

There was nothing of interest to me here, so I moved on again. The first room on the back wall was more interesting. There was a single desk in this one on the opposite side of the room to the door. There were shelves that stretched almost all the way around the room, filled with buff files.

I picked one at random. From what I could see, it appeared to be some kind of dossier on a person of interest to the Institute. At the front was a fading colour photo of someone I didn't recognise steering a small boat, wearing a sailor's hat and fisherman's smock, clearly blown up from a much smaller original. Someone had drawn a large X from corner to corner using a blue felt-tip pen, which struck me as a slightly sinister touch.

Following this, there were a number of yellowing press clippings (mostly, it seemed, about someone called Dominic Bilston, so I assumed that was who it was in the picture), along with some scribbled notes describing his movements at various times, dating between 1973 and 1987. I hadn't realised that the Institute had been around for that long.

I took out my phone and googled the name. It turned out that he was some kind of entrepreneur with interests in sub-Saharan Africa, who had died in a car accident in 1987 on the road out of Mombasa. There were, apparently, no suspicious circumstances, however.

Right.

By this time, my eyes had readjusted to the pale light and I noticed that there was another file sitting on the desk in front of the chair. There was a Post-It note stuck on the outside with the words 'FAO Gowers' scrawled on it, but the name didn't mean anything to me. I picked it up and looked inside. Once again, there was a picture at the front, much more recent than the one of Dominic Bilston, and this time it hadn't been crossed out.

But the man in the picture still wasn't anyone I recognised, even though there was something just slightly familiar about him. However, when I turned it over and began looking at the rest of the file, those hairs on the back of my neck went into serious action again. One name kept coming up again and again: Sergei Kravchenko.

I didn't know what our Sergei's surname was, but what clinched it was a clipping showing both him and his younger brother Maxim being presented with the Litvinchuk Medallion of Honour at some university in Belarus. I remembered the story of Sergei and Maxim being the only pair of brothers ever to be jointly awarded that prize. I also remembered that I'd actually seen Maxim once, because I'd been there when the Gretzkys had run him over and bundled him into the back of their limousine. And that, of course, explained why the photo of Sergei looked familiar.

But why did they have a whole file on him? Was it just because the Institute were after the Vavasor papers? More to the point, why did they have a whole roomful of files on other people? Was there some kind of hit list going on here?

If so, was I in there somewhere? Was Dorothy?

Assuming that they were in alphabetical order, I hunted down where Chan, Dorothy might be. She wasn't there. Neither was Winscombe, Tom. Well, that was something. I guess they hadn't sussed out who we were yet.

I tucked Sergei's file under my arm and left the room. The door in the middle of the back wall led into a tiny broom cupboard, and a moment's glance told me it wasn't worth bothering with. The next room was completely empty, so I was halfway round now.

My phone vibrated and another WhatsApp message from Dorothy appeared.

How's everyone doing? she said. *I've got nothing.*

Nothing yet, said Arkady.

Nothing here either, said Balvinder.

Found a file on Sergei, I said.

Cool, said Dorothy. *Anything on the Vavasor papers in it?*

Hold on, I said. In my haste to find out if we were on the Institute's hit list, I hadn't actually checked that. I flipped through the file again, but the only Vavasor I came across was Isaac, the twins' younger brother, and even that was only in the context of him employing Sergei as a security consultant.

Nope, I said. *Nothing there.*

OK, keep looking everyone, said Dorothy.

We all replied with a series of 'thumbs up' emojis and I put the phone back in my pocket. The next two doors both opened into the same room, which was laid out as a kind of lecture theatre, with a series of desks arranged in rows. There was a screen at the right-hand end of the room. So no Vavasor papers in here either. I was doubtful about the boardroom, too, so that left the room next to the lecture theatre. I opened the door into a pitch black cubby hole. I shone my torch in there and it turned out to be full of what looked like extremely expensive audio visual equipment, some of which appeared to be feeding into the lecture room next door and some into the boardroom.

I stepped out into the corridor again and it was at this point that I noticed that the door to the boardroom was now open. I was sure I hadn't touched it yet, but there it was, open all the way. I thought back over the time I'd spent up on the second floor, and despite what the neck hairs might have been screaming at me, I hadn't heard anyone other than myself. I was sure of this.

Maybe it was just the wind or something.

As if there was a gale blowing up here. I licked my finger and held it up. No, there wasn't even so much as a mild breeze, and that door looked pretty solid.

In the absence of any form of weapon, I rolled up Sergei's dossier and grasped it firmly in my right hand. I gritted my teeth and walked over to the open doorway into the boardroom. I hesitated for a moment or two on the threshold, trying to see if I could hear anything over the industrial din that my heart was currently banging out, but I couldn't make anything out at all.

I stepped into the room. Then, with a soft click, the door shut gently behind me.

Chapter 4

The room was in complete darkness. What windows there were at the front of the building must have been covered by thick blackout curtains. I blundered back towards the door I'd just come through, but I only managed to take half a step before I ran into something that I quickly realised was another human being. Then a pair of firm hands pushed me in the chest, propelling me backwards into the room. I struck out with Sergei's dossier, hoping that the full weight of a medium-sized sheaf of A4 might have some impact on my attacker. But all that happened was that the file was grabbed neatly out of my hand and I was left flailing wildly at thin air and struggling to maintain my balance.

As quietly as I could, I dropped down onto the floor. My eyes were still struggling to adjust to the lack of light, and I decided I was less likely to trip over something if I was on my hands and knees. The fact that it was so dark gave me a brief moment of hope. If I couldn't see whoever was in the room with me, then presumably they couldn't see me either. Unless they were wearing some kind of night-vision goggles, of course, in which case I was completely screwed. But that would mean they were professionals, which meant I was screwed anyway.

This was the boardroom, which meant that if it was like any of the other boardrooms I'd given presentations in during my PR career, there would be a big, solid, chunky table in the middle of it. Perfect for hiding under. So I wriggled my body over to where I thought the table might be. Sure enough, I very soon encountered solid table legs. I tucked myself in and considered my next move.

I had to contact the others. Arkady would have this guy for breakfast. I carefully took my phone out, reckoning I could somehow send a WhatsApp message to the rest of the group without the light of the screen drawing myself to anyone's attention. In this, however, I was one hundred per cent wrong, because before I'd completed a single word, a hefty boot had kicked the phone out of my hand. Then a firm hand gripped hold of my right ankle and dragged me out from underneath the table.

Shit.

I kicked out with my left foot but failed to connect with anything.

'Let go of me,' I said. There was no response. 'You're going to regret this, you know,' I added, although the bravado didn't quite come off as convincingly as I'd hoped. 'Seriously. I have friends.'

There was a gentle, patronising laugh in the darkness, and I felt mildly insulted. Then I became aware that someone was bending down near my face. Hair that for some reason I just knew was blonde brushed briefly against my cheek and a finger was pressed to my lips.

'But—' I began.

The next thing I knew, something was being sprayed in my face, and I had a brief flashback to a time when I was thrown out of a taxi heading into Kent after springing Benjamin Unsworth from St Jude's Hospital following his poisoning by goons associated with Channellia. I could feel the hairs on the back of my neck saying, 'There! What did I tell you? I *knew* there was trouble about. And what did you do about it?'

'Nothing,' I muttered to myself as I lost consciousness. 'Nothing at all.'

I was sitting in a chair in the middle of a room. The morning sun struggled weakly against the net curtains and decided in the end not to bother. Through the curtains I could see that there were bars across the window. The walls were papered with one of those flowery designs that looked like faces if you squinted and tilted your head to one side. Apart from the chair I was perched on, the only other items of furniture in the room were a table and another chair over by the window, an iron-framed bed against the wall behind me and a washbasin in the corner.

My hands were secured behind my back with what seemed to be one of those disposable plastic cable ties. I suddenly felt sick, so I leaned over to one side and vomited. Someone had clearly anticipated this eventuality, because there was a bucket there waiting for me.

The door opened and a familiar person walked in, dumping a briefcase on the table. She sniffed the air, walked over to me without saying a word, picked up the vomit bucket and carried it out of the room. Then she came back in, carrying a glass of water and a can of air freshener, which she proceeded to spray liberally around the room. Then she placed the can on the table and brought the glass of water over to me, placing it against my lips. I took a few greedy sips, but not quite enough to clear the unpleasant taste in my mouth.

'Matheson,' I said.

'Sweetheart,' she said, stepping back and standing a few feet away from me with her arms folded. 'How lovely to see you again, and so soon after our last little encounter.' She was wearing a plain white blouse and a pair of tightly cut jeans. For all I knew, they could have been bought at Primark, but she managed to make them look like a million dollars. I tried very hard not to be taken in. I wasn't in any mood for exchanging pleasantries.

'Yeah, whatever,' I said. 'Look, could you possibly undo this tie. It's quite painful, you know.'

'Depends on whether you're going to make a bid for freedom, Tom. I'd hate you to run out on me now we're in a position to do business.'

'I have no idea what you mean.'

'No, I don't suppose you do.'

I realised there was something nagging at the back of my mind.

'Where's Dorothy?' I said.

'Ah, I was wondering when you were going to ask about her,' she said.

'Well?'

'Well what?'

'Well, where is she?'

'All in good time, Tom. All in good time.'

'That's not terribly reassuring,' I said.

'No, I can understand that,' she said stepping closer and squatting down until her eyes were level with mine. 'But sweetheart,' she added, lifting my face up to hers with a single finger under my chin, 'I'm not the one who's just been caught burgling an organisation like the Institute for Progress and Development. Am I?'

'Wrists, please,' I said. 'You know I'm not going to go anywhere until I know what's happened to Dorothy.'

Matheson stood up and walked behind me. I felt a strong, sharp, upward movement against my restraints and then the tie fell off and my hands were free. I rubbed them together and wondered briefly about whether I needed to keep my word. It surely wouldn't be too hard to make a run for it.

But when I tried to stand up, it turned out that my legs were also constrained by ties, and that – combined with the fact that I still hadn't fully recovered from whatever it was they'd used to put me out – meant that I only managed a single step forward before I collapsed into an awkward face plant on the floor with the chair now lying playfully on top of me. I managed to haul myself into a kneeling position and just about succeeded in pushing the chair upright again. Then with great difficulty, I hauled myself painfully slowly back into the chair, all the time being watched by Helen Matheson with a sardonic smile on her face.

'Nice try, Tom,' she said.

'I just wanted to stretch my legs,' I said.

'Of course you did, darling. Of course you did. But can I just explain one thing?' She gave an exaggerated nod and raised an eyebrow as if to suggest that I should do likewise.

I gave a weary nod by way of response. 'Good, Tom. Now, watch.' She walked over to the door and opened it. Then she disappeared through the doorway and came back leading one of the largest men I'd ever seen by the elbow.

I'd always been impressed by Arkady's muscular physique but he was a mere bantamweight compared to this guy, who seemed to have been constructed from the discarded contents of two or possibly three horses. The most disturbing part of it was his head, which was a normal-sized human one, which meant that it was roughly half the width of his neck. It didn't look right at all. I wondered if he was the brute who had attacked me in the boardroom in the early hours. Given that I had survived the experience, it was unlikely.

'This is Brett,' said Matheson by way of introduction. 'He'll be your guard while you're with me.'

Brett amplified this with a grunt and an attempt at a smile.

'Cool,' I said. There probably wasn't a lot going on inside that weird little head of Brett's, but it was almost certainly enough to coordinate his motor skills sufficiently to cause me a lot of pain if I tried to make any trouble.

Matheson smiled at me as if to say 'OK?' and ushered Brett out of the room again. 'Hands behind your back, please.'

'What? But you've got that lump Brett guarding me.'

'I do. But someone needs to be taught a lesson. Hands.'

I thrust them behind me once more, and Matheson did the honours with a new cable tie. She made it tighter this time, which struck me as an unnecessarily vindictive touch.

'Where's Dorothy?' I said.

'Let's leave Dorothy out of this for the moment,' said Matheson, coming out from behind me. 'I want to talk about you.'

Helen Matheson had lately become a recurring feature of my life, and one that I had very mixed feelings about. I'd first encountered her when I'd gone to the hospital to find Benjamin Unsworth, and it was she who had thrown me out of the taxi we ended up sharing with him on the way to an unknown destination in Kent. Then she'd shown up following Dorothy and myself when we were trying to retrieve a dangerous PhD thesis belonging to Dorothy's business partner Ali's girlfriend Patrice from the Fractal Monks.

I still wasn't entirely sure who it was that Matheson worked for or if she was completely freelance. I'd originally assumed that she was part of the secret services, judging from the way she'd used a weapon to help us escape from the hospital, but she'd claimed that her department had been privatised and she'd actually work for anyone who would pay her. However, she didn't appear to have much in the way of a decent budget for hiring staff with any great staying power.

Of the ones I'd encountered, there was, for example, Matt Blank, who'd gone undercover for her as PR adviser for the Institute for Progress and Development's cryptocurrency operation and had paid for it with his life when they'd rumbled him. And there was also Bancroft, who had been killed by the Fractal Monks shortly after he'd tried to steal a laptop that I'd retrieved from the underwater wreckage of Channellia.

And of course, Benjamin Unsworth himself had ended up working for her briefly, which was entirely consistent with his record of poor choices of employment. He'd previously been an intern for Hilary van Beek, the Vavasor twins' biographer's publisher, before she got strangled with a spline by Rufus Fairbanks in an attempt to cover up his connections with the Belarusian mafia. Benjamin was last seen acting as general dogsbody and alpaca wrangler for Margot Evercreech, the prominent Vavasorologist.

No, Helen Matheson didn't have a great record as an employer, as most of the people she took on turned out to be either hopeless or dead. Or both.

'How would you like to work for me?' she said.

'I'm sorry?'

'How would you like to work for me, Tom?'

'I'm a bit busy at the moment,' I said.

Matheson executed an elaborate pantomime of standing back, looking me all over and shaking her head with exaggerated slowness.

'Really?' she said.

I sighed. 'You know what I mean.'

'I'm not sure I do. You look distinctly unencumbered by the burdens of work at the moment.'

'Look, what I meant was, as soon as you decide to let me go, I've got plenty of other stuff to be getting on with.'

'Do tell.'

'Well, there's…' The trouble was, I couldn't say anything about it without giving something away. 'There's just lots of stuff.'

Matheson gave a supercilious smile. 'So are you intending to go breaking into more buildings, then? Is this where your career is taking you now?'

'No, of course not!'

'Tom, what were you doing there?' she said suddenly.

'What?'

'What were you doing in the Institute?'

'I could ask you the same question,' I said.

'Yes, but I'm the one in charge here, Tom. My gaff, my rules. So let's try and break it down a bit. There are two reasons for breaking into an organisation such as the Institute for Progress and Development – espionage or theft. Now you don't strike me as the sort of person who would be interested in spying on people, so I'm thinking you must have been looking to steal something. And it must be something sufficiently important to you that you were prepared to risk being caught. Am I right?'

I said nothing.

'I'll take that as a yes, then,' she said. 'But what could that be? Knowing what I know about you, the only possible thing I can think of is – oh god, it's the Vavasor papers, isn't it?'

'I—' I stopped myself from saying anything but it was too late. A gleam had come into Matheson's eye.

'Well, well,' she said. 'What is it about those papers that gets everyone so excited?'

I simply shrugged.

'More to the point, what is it about them that gets *you* so excited, Tom?'

I still didn't trust myself to answer her. Matheson moved closer to me.

'You see,' she said, 'I have this odd little feeling that you couldn't actually give a monkey's one way or another about the Vavasors. It's all about Dorothy, isn't it?'

'Leave her out of this,' I said. 'Where is she, anyway? I'm not going to do anything until you at least tell me if she's still alive.'

'Oh, she's still alive,' said Matheson, casually. 'I really don't get what you see in her, though. Is she really worth it, Tom? Why would a semi-intelligent man like you allow yourself to risk everything by being dragged into a wild goose chase after some stupid papers that you haven't even the slightest interest in?'

At this point, Matheson shocked me by stepping right up and straddling me, leaning in to whisper in my ear.

'Is the sex good, Tom?' she said. 'Is that it?'

'Get off me,' I said, trying to push her away with my body by wriggling and ending up making a distinctly ambiguous manoeuvre that for some reason she found terribly funny. Then she stood up, walked around to my back and placed her left hand over my chest for a few seconds.

'I'd watch that heartbeat of yours,' she said. 'Bit fast for this time of day.'

'Where is she?' I snapped.

Matheson circled round and stood in front of me, her arms folded once more.

'Dorothy is safe,' she said. 'For now. But her continued safety may depend on how willing you are to cooperate, Tom.'

'What do you mean?'

'I'm sure I don't need to elaborate.'

'Are you trying to blackmail me?'

'God, you can sound so pompous, Tom. Are you going to be like this all the time?'

'Oh, piss off.'

Matheson slapped me across the face in response to this.

'Ow!' I yelped.

'Well, behave yourself,' she said. 'I can't be doing with insubordination.'

I sat in silence for a few moments.

'When do I get breakfast?' I said.

'When we've finished talking,' she said, brightly.

'So this is how it goes,' said Matheson, pacing the room. 'A client of mine is having some issues with the Belarusian mafia.'

'I'm sorry?' I said. I already didn't like the sound of this.

'The Belarusian mafia, Tom. Every country has a mafia.'

'I know.'

'And this one's in Belarus.'

'Right.' This was probably the point where I should have mentioned my little contretemps with the Gretzky gang, because this would clearly invalidate my involvement in any further Belarusian adventures. Looking back on it, however, that was exactly the reason I was reluctant to mention it. If I was to save Dorothy, I would have to go through with this, however bad it got.

'Now, the thing is, there's a person of interest to my client who has gone missing in Belarus while working for one of the mafia gangs and we need to know whether he is dead or alive. Your job is to find out which.'

'Right. And why me, exactly?'

'I need someone who has no prior association with me. Also, despite a superficially high level of general incompetence, you appear to have a remarkable ability to get out of situations that would prove fatal to any normal human. It's a skill that shouldn't be underestimated.'

I wasn't sure if this was a compliment or not. I took it as such, but I also had to be honest.

'You should really be talking to Dorothy, you know,' I said. 'She's the one who devises all the plans, not me.'

'Oh Tom,' said Matheson, walking over to me, bending down and taking my head in both hands, 'you are way too modest, my darling.' Then she stood up and added, 'and besides, I can't use her.' She went over to her briefcase and took out a single sheet of paper, which she held up in front of me. There was a picture of a man with long hair and a wild, straggly, greyish beard, with the name 'Rory Milford, Faculty of Mathematics' printed underneath.

'Who's this?' I said.

'The man you're going to impersonate, Tom.'

Chapter 5

'OK, Matheson,' I said, 'I can see a number of problems here. First of all, what if the real Rory Milford turns up?'

'He won't.'

'Why not?'

'Because he's currently sitting in a room elsewhere in this very building, wondering what the hell's going on.'

'Good grief. You're not planning to kill him, are you?'

'Good heavens, sweetheart, I'm not a monster, you know. Once you're safely back, we'll let him go again. I guess we'll have to keep him occupied with sudoku and things, but the budget should cover that.'

'Right,' I said. 'OK, secondly, I don't speak Belarusian.'

'Neither, as far as we know, does Dr Milford.'

'As far as you know?'

'Yes.'

'All right, let's park that to one side for a moment. Thirdly, he doesn't even look like me. He's older, too.'

'We can pretend you've just shaved your beard off. And in actual fact, he's not that much older than you. He's just got an unhealthy lifestyle.'

'Yeah, well, that's as may be. Fourthly, and this is quite a biggie, I'm not a mathematician. That, once again, is Dorothy's department.'

'That won't be a problem.' She went back to her briefcase and took out a large sheaf of paper. 'I've got copies of all his papers for you to read,' she said, waving them at me. 'You can give them a scan through on the flight over.'

'There's no way I'll be able to understand that,' I said.

'No, you're right. There probably isn't. It's the weakest part of the plan, but try as I could, I couldn't lay my hands on a lookalike postdoc mathematician at short notice, Tom. You just can't get the staff these days.'

'I'm going to die, aren't I?'

'Only if you cock the mission up, Tom. All you need to do is introduce yourself saying you're the chaos theory expert from Cambridge, and—'

'Chaos theory?' Oh god, it was getting worse by the minute. The only thing I knew about chaos theory was that it was related to fractals and the only thing I knew about fractals is that they made my brain ache.

'Yes, Tom. I thought you'd be interested in that, given that chaos is a bit of a theme with you.'

'Not funny, Matheson.'

'No, I don't suppose it is. Anyway, you ingratiate yourself with our merry band of mafiosi, find out what's happened to my man, then do a bunk when no one's looking. Should be back in time for tea.'

'I really am going to die,' I said. 'I don't want to die quite yet. And in any case, what do they need a chaos theory expert for?'

'Very good question, Tom. And the answer is we don't know yet, as Dr Milford's being singularly uncooperative. But he'll talk eventually.'

'I bet he will. So how did you find out about him?'

'Oh, one of my associates at the university got a tip-off last week that the mob had engaged him, so we put a plan together *tout de suite*.'

'And I just happened to make myself available, did I?'

'You did.'

'But why all the shenanigans at the Institute for Progress and Development?' I said. 'Why didn't you just phone up and ask me?'

'Would you have said yes?'

'Of course not.'

'Well, there's your answer, Tom. Besides, you weren't even in the frame for the job until the early hours of this morning.'

Not for the first time lately, I had a feeling that I was getting way out of my depth here, and having spent my entire life up to this point splashing happily around in the shallow end, I was finding it hard to adjust.

'Can you run that past me again?' I said. 'I think I might have missed something.'

'Up until my associate at the Institute happened to bump into you a few hours ago, we had a bit of a casting problem. None of my usual players looked remotely like our man Milford, and we were wondering what on earth we were going to do.'

'Did you just refer to your associate at the Institute?'

'We have interests there too, Tom.'

An awful possibility suddenly presented itself. 'Are they clients of yours?' I said.

'Good heavens, no,' said Matheson. 'But I have to say that your botched little raid may well turn out to be quite a boon, if only to distract their attention from the fact that one of their researchers is actually one of ours.'

'Bloody hell,' I said.

'Well, you shouldn't be too surprised, Tom. Remember we've had assets embedded with that lot before now.'

I suddenly had a flashback to the first time I'd come across the Institute, when I'd witnessed them kill her man Matt Blank in cold blood. Despite the way he'd treated me, I hoped that her new guy would have a better outcome.

'Ah,' I said.

'Yes, no one's noticed anything unusual about him yet. There are some frighteningly clever people working there, but they're also spectacularly dim when it comes to their own security. As you no doubt noticed when you managed to break in there.'

'So was your guy there already?'

'Well, not exactly. But he received an alert of some unusual activity and he decided to head in to see what was going on. Better for him to be looking into it than one of their regular guys after all.'

'But I thought we'd switched all the alarms off,' I said.

'All the regular systems, yes. But my man installed a couple of his own. Just to keep one step ahead.'

'Good grief,' I said, although part of me was secretly delighted that Dorothy had actually missed something. She was sometimes just a little too perfect for my liking.

'Anyway,' said Matheson, 'I get the backup team together and my man goes in, sneaking up the back stairs without anyone noticing and you probably know most of the rest. By the way, we were terribly impressed with the work your young lad did on the basement door. I could have made use of him.'

'Well, you're not having h— oh god is he OK? And the others?' In my concern for Dorothy, I'd completely forgotten about the rest of the team.

Matheson paused for a moment before replying. 'All will be revealed in due course, Tom,' was all she said. I didn't find this remotely reassuring. Then she suddenly clasped her hands together and said, 'Breakfast!' as if it was the most natural thing in the world.

It was good to have my hands freed again, but for once I didn't feel like eating. I was being asked to undertake a difficult and almost certainly suicidal mission on behalf of a deranged lunatic who was apparently holding Dorothy hostage. For all I knew, the rest of the team were dead or at the very least in serious trouble. Also, I'd just remembered something else important.

'Where's my phone?' I said.

'All in good time,' said Matheson, sitting at the table under the window and buttering a slice of toast. 'Are you sure you won't have anything? The marmalade is to die for. Lovely lady down the road from here makes it herself.'

I shook my head. 'Where are we anyway?'

'A safe house.'

'Safe for you or safe for me?'

'Both of us, Tom. But, yes, mainly me.'

'How many people do you have working for you?'

'Ooh, that would be telling,' said Matheson, stuffing the toast in her mouth. 'Mmmm, that is SO good,' she added.

'How many of them have died?' I said.

Matheson patted her lips with a white serviette and adopted a serious face. 'Tom,' she said, 'this is a risky line of work. You've seen what we do. You've seen the people we have to deal with. Coffee?' She gestured towards the cafetière that she'd brought in with the rest of the breakfast things.

'Yeah, I guess so. Black. No sugar.'

'Of course.'

Matheson pressed the cafetière's plunger and then poured the contents into two mugs, handing one to me. I took a few sips and as the caffeine began to take its effect I started to feel a little better.

'You're not going to tie me up again, are you?' I said.

'No, Tom. I think we have an understanding now, don't we?'

'I want to see Dorothy.'

Matheson pursed her lips. 'I'm not sure that would be a good idea right now.'

'Why not?'

'Let's talk a little more about your – well, shall we call it a mission? I do think "mission" sounds awfully exciting.'

'It sounds terrifying.'

'Well, that too, Tom. That too.' She paused and then continued. 'So, you're booked on a flight out of Heathrow to Minsk tomorrow morning. You have Rory Milford's passport and a stack of Belarusian roubles, which you shouldn't actually need to use unless in emergencies. You'll be met at the airport, so remember to look out for a sign with your new name on it.'

'What if something goes wrong? Can I call you?'

'No, Tom. If you call me, there's a very good chance I will deny all knowledge of you.'

'Even if it's something really bad and I need to be rescued?'

'Sorry, Tom. We don't do rescues for deniable assets.'

'For what?'

'We don't rescue anyone who is classified as a deniable asset. And by an amazing coincidence, that's also your official job title now. Deniable Asset, Grade Two.' She gave me an encouraging smile, as if this was good news.

'Jesus. What do I have to do to get promoted to Grade One?'

'Stay alive, darling.'

'Do I get a contract of employment?'

Matheson sighed. 'Tom, did you hear what I just said? Specifically the word "deniable." Deniable assets don't get a paper trail.'

'How much do I get paid?'

'We'll see what's left over after all the expenses have been taken care of.'

'So not even minimum wage?'

'It's a tough business, darling.'

'So not only am I going to die, I'm going to die alone in a foreign country with no help forthcoming and no money to leave to anyone.'

'To be fair, Tom, I don't think Dorothy needs your money.'

'It's the principle that counts.'

'Well, there's a turn-up. I really didn't have you down as a man of principle.'

'Oh, stop it. Let's get this over. What else do I need to do to get Dorothy released?'

By way of response, Matheson spread her arms wide. 'Nothing, Tom!' she said with a wild-eyed grin. 'Nothing at all.'

I was clearly missing something important.

'What do you mean?'

'I mean we're not holding her. Any of them. Neither Dorothy nor the other two.'

'Two?' I said.

Matheson frowned. 'Yes, two. Why did you say that?'

'No reason at all,' I said. So what had happened to Arkady, then?

'Tom, did we miss someone? I need to know. This is important.'

'No, no. There were just the four of us.' I hoped I wasn't blushing. I was a very bad liar.

Matheson stared at me for a while, as if wondering whether or not to pursue this line of discussion any further.

'Going back to what you said about not holding them,' I said, 'what exactly did you mean by that?'

'I mean, Tom, that we let everyone go after we closed down the raid,' she said. 'My budget doesn't stretch to bed and breakfast for an entire gang of hungry burglars, you know. I mean, the alternative was to kill them, but that seemed a tad messy. So no, we let them go.'

'But that means...' I tailed off.

'Let me help you out, here,' said Matheson. 'You're thinking it means that I've got no hold over you any more and that you can go back to your Dorothy and everything's going to be happy ever after. Well, everything apart from my operation, which is going to be significantly compromised by not having anyone involved who actually looks like the man whose identity we're trying to fake. But I expect you're not too bothered by that.'

'No, I'm not,' I said. 'Right, then. It looks as if our discussions are at an end. Can I go now, please?'

'Of course.' Matheson waved a hand towards the door. I stared at her. Something wasn't right at all about this. I could smell a rat and it wasn't just any old rat. It was a giant man-sized plague rat whose maggot-infested corpse was malodorously decomposing right under my nose.

'Can I have my phone back now, please?' I said.

'Of course,' said Matheson, reaching into her briefcase and tossing it to me. I caught it and switched it on just to make sure everything was OK.

'You might want to check a few things before you call Dorothy to say you're on your way,' she added.

'What?'

'Oh, just make sure you're happy with your last few WhatsApps. That sort of thing.'

'Matheson,' I said. 'What have you done?'

The first few new interactions on the 'Raiders of the Lost Institute' WhatsApp group were innocuous enough. They started about half an hour or so after our previous exchange of messages about finding Sergei's file.

03:14[Dorothy] *Progress, anyone?*
03:15[Arkady] *Nothing*
03:15[Katya] *Zilch*

03:17[Dorothy] *Tom?*

03:19[Dorothy] *Hello, Tom? Wake up!*

03:21[Balvinder] *Think I heard something upstairs.*

03:21[Dorothy] *What sort of something?*

03:21[Balvinder] *Dunno. I'm going to take a look.*

03:21[Dorothy] *STAY RIGHT THERE.*

03:22[Katya] *Too late, he's gone off*

03:22[Dorothy] *WELL GET HIM BACK*

03:24[Katya] *OK I got Bal. Tom's not here.*

03:24[Dorothy] *WHAT*

03:24[Balvinder] *What she said. All the rooms are dark.*

03:26[Katya] *OK we're coming down. Something's not right.*

03:28[Dorothy] *WAIT THERE'S SOMEONE AT THE FRONT DOOR*

03:28[Dorothy] *WHATS HAPPENING*

03:29[Dorothy] *ARKADY*

03:30[Katya] *Dorothy, what's going on? I can hear someone coming down behind us.*

03:31[Dorothy] *SHIT*

03:31[Balvinder] *What'sasdhasd;io*

There was a break after this until a new thread started:

04:02[Dorothy] *Tom? Where are you?*

04:03[Katya] *Дзе ты?*

Clearly, Katya had switched to Belarusian in panic. Which wasn't a good sign.

04:03[Dorothy] *Are you OK? Where are you?*

04:05[Dorothy] *Come on, Tom, joke's over. We're OK. Got Katya and Bal here. Bit shaken up but OK. Need to know you're safe.*

When I saw the next message, ten minutes later, my blood ran cold.

04:15[Tom] *Hi everyone. Sorry about all this.*

There was no way I could have written that. The only possible conclusion I could draw was that it was Matheson. At first, I was amused that she was so wrapped up in the idea of pretending to be me that she hadn't bothered to read the earlier messages that indicated Arkady's presence in the team. However, my amusement very quickly gave way to panic. How much worse could this get?

04:16[Dorothy] *Tom! You're alive. Thank god for that.*
04:16[Tom] *Yeah well. Like I said, sorry for dropping you in it like this.*
04:16[Dorothy] *What do you mean*
04:17[Tom] *Haven't been entirely straight with you guys.*
Christ, NOW what?
04:17[Dorothy] *WHAT*
04:17[Tom] *Had a call from Matheson the other day. Remember her?*
04:17[Dorothy] *I REMEMBER HER*
04:17[Tom] *So, yeah, she asked if I could help her with a little project and I couldn't really say no*
04:18[Dorothy] *YOU SO COULD TOM*
04:18[Tom] *Well, yeah maybe. But I didn't want to cos it sounded more fun than the sodding Vavasors.*
04:18[Dorothy] *WHAT*
04:18[Tom] *I guess I should have told you in person. Well, I did try to tell you but you wouldn't listen.*

04:19[Dorothy] *NOT TRUE TOM NOT TRUE*
04:19[Tom] *Not worth discussing at four o'clock in the morning*
04:20[Dorothy] *BLOODY IS*
04:20[Tom] *Also*
04:20[Dorothy] *ALSO WHAT*

At this point the conversation stopped and I couldn't work out why until I noticed that it had moved to the private channel between myself and Dorothy. And that's where things took a real turn for the worse.

04:21[Tom] *Look I didn't want to say this in front of the others but you and me it's not working is it?*
04:21[Dorothy] *IT'S NOT WORKING RIGHT NOW THAT'S FOR SURE*
04:22[Tom] *Please stop caps locking me.*
04:22[Dorothy] *SOD OFF*
04:22[Tom] *So maybe we should take a break or something*
04:22[Dorothy] *WHY NOW TOM*
04:22[Tom] *Why not*
04:22[Dorothy] *ARE YOU WITH HER NOW*
04:22[Tom] *Maybe I am maybe I'm not*

'Matheson,' I said, making a brief and undignified attempt to stand up before remembering that my lower half was still attached to the chair and sitting heavily back down again. 'What the fuck have you done?'

'Think of it as me helping you out of a bad place,' said Matheson.

'I don't need any fucking help.'

'Please don't swear, sweetheart. It's unnecessary and vulgar.'

'I don't fucking care. You've got absolutely no right to meddle in my private life.'

'Private life? The only reason, Tom, that Dorothy is still alive is because I like you. You're absolutely right that she's the brains of your little outfit. But over the last few months, that woman of yours has caused me so much aggravation that if she were here now, I would struggle to restrain myself from throttling her. I don't even know what you see in each other.'

'That's between us.'

'Well, not for much longer.'

'We'll see about that.'

'Really? Scroll on, darling. Scroll on.'

I did, but when it came to the point where Matheson was describing to Dorothy the full details of the new and comprehensive sexual relationship that she and I had been nurturing these last few weeks, I lost my temper completely and hurled the phone across the room at her, accompanied by an anguished howl that expressed my feelings far more eloquently than any words I could come up with. It missed, crashing into the wall beyond her.

Matheson got up from her chair, bent down to pick the phone up, walked over and handed it to me, all without a single word being spoken. I looked at it. There was a crack in one corner of the screen, but apart from that it appeared to be still working.

'You should be careful with that. Do you know how much they cost to replace?'

I said nothing. I quickly scanned the rest of the conversation between Dorothy and Matheson-as-myself, which left very little doubt as to the level of undiluted naked fury that it had aroused in my almost certainly now ex-girlfriend.

I stared at the phone, willing the conversation to somehow disappear. But it didn't.

I checked the time. It was just after eight o'clock in the morning. Well, there wasn't going to be any better time to do this, however long I waited. I pressed the call button, expecting it to go straight to voicemail. Instead, Dorothy answered.

'What,' she said.

'I… look,' I said, 'hear me out for a moment and please don't hang up. I'll understand if you're upset and I know I would be in your position but it's really not what it looks like at all.'

Oh god, that was terrible. Dorothy said nothing. I could hear her breathing on the other end of the line.

'You see,' I said, 'it was Matheson who wrote all that stuff, not me, you see. It's not true at all.'

Dorothy still didn't say anything.

'None of it is true. She's kidnapped me and she's tried to make me agree to work for her and I said I would because I thought she was going to hurt you and—'

'Is she with you now?' said Dorothy.

'Well, yes, but—'

'Goodbye, Tom.'

The line went dead before I could say anything else. I put the phone in my pocket and glowered at Matheson.

'I take it that didn't go well,' she said, breezily.

'No it fucking didn't.'

66

'In which case, I hate to break this to you, but I don't think your Dorothy was as invested in your relationship as you may have hoped.'

'What do you mean?'

'Well, she does seem to have given up rather easily.'

'I despise you.'

'That's a bit harsh, Tom.'

'Is it? I don't think so.' The truth was, I'd never heard Dorothy sound so lost or defeated before.

'Well, I'm sure you'll get over it,' said Matheson. 'Plenty more fish and all that.'

I said nothing.

'So are we good, then?' she added.

'Sorry, what?'

'Are we good? You're still OK for the job?'

'Are you kidding? Give me one reason why I should ever want to have anything to do with you now or ever again.'

Matheson essayed a sympathetic smile. 'Look,' she said. 'I'm sorry it didn't work out with Dorothy. You're bound to feel sad for a while. Would you like a moment in private?'

I didn't say anything.

'No?' said Matheson. 'All right then. Your flight is at half past eleven tomorrow morning, so we'll need to leave here after breakfast.'

I stared at her. 'Did you even hear a single word I said? I'm not doing it. Forget your stupid mission. I'm not interested.'

'But I need you.'

'Well, you should have thought about that before you went and interfered, shouldn't you?'

'I thought I was doing you a favour.'

'Oh, please.'

'Are you sure there's nothing I can do to convince you? I mean, something like this could help take your mind off things for a while.'

'Are you mad?' I said.

'I don't think so. I'm just trying to help you deal with what must be a really difficult situation for you.'

'Which you caused, Matheson.'

Her response to this was a shrug.

'Look,' I said, pointing down to my ankles. 'Just untie me and show me which way to get back home. I'm not even going to ask for the bus fare, OK? Just let me go.'

'I don't think I can do that, Tom.'

'What? Why ever not?'

'Well, I've just told you all my plans for one thing.'

'I promise I won't tell anyone. I mean, really I won't.'

Matheson folded her arms and tilted her head on one side as if giving something serious consideration.

'I believe you, Tom. But that may not be enough for my client.'

'Oh, come on. This is getting silly. Just let me go. Please.'

'I don't know, Tom. I don't know at all.' She went silent for a few seconds. 'Oh, Tom,' she said eventually. 'You're making this so difficult. I like you. I don't want anything to happen to you.'

'Well, it's not going to, is it?'

'Isn't it?' said Matheson, reaching across to her briefcase. She pulled out a compact revolver, along with what I assumed was ammunition. She fiddled around with it for a while and then held it up and pointed it at me.

'Jesus,' I said. 'Can you point that somewhere else?'

'Sorry, Tom.'

'This isn't fair. None of it is. I don't deserve to die.'

'No, Tom, you don't.' Matheson put the gun down on the table. For a moment I tried to work out whether or not I could launch myself across the room and grab it, even if my legs were still tied to the chair. But it was hopeless. I didn't stand a chance. 'Come on,' she said. 'Don't make this difficult for me. I've had a tiring day.'

I thought hard about this. There was every chance that, in the end, Matheson would get bored and shoot me. I'd seen her shoot someone without a moment's thought when we snuck Benjamin Unsworth out of hospital that time, so I was under no illusions about her willingness to kill someone if she needed to.

'I'm going to die whatever happens, aren't I?' I said. 'I'll either be dead in a ditch somewhere in Belarus or buried under a nice new patio out the back here.'

'Ooh, that's a nice idea,' said Matheson, clapping her hands. 'It'd be lovely for a barbecue.'

'Don't. It's not clever.'

'No, I suppose the weather's too unpredictable. Look, are you going to work for me or aren't you? Because if you are, we need to get all the arrangements finalised.'

'Do I have a choice?'

'I think you know the answer to that, Tom.'

There was a long silence between us, in the course of which we seemed to come to some kind of understanding. Matheson nodded, then unloaded her gun again and put it away in her briefcase.

'OK, we have some paperwork to go through,' she said. 'But first of all, I need to ask you about this,' she said. And then something happened that changed my perspective on the entire project. She took out a buff folder and passed it across to me, with a curious expression on her face. I recognised the folder immediately. The last time I'd had it in my hand, I was trying to use it to swat away the man who was attacking me in the boardroom at the Institute.

'How did you come to have this in your possession?' she said.

'I… just happened to find it lying around,' I said. 'Thought it looked interesting.'

'It's very interesting. To me, as well as to you, as it happens.'

'Sorry?' I said. 'Why's that?'

'This,' said Matheson, 'is Sergei Kravchenko. He's the guy you're going to go and find for me.'

Chapter 6

I spent the rest of the morning reading and re-reading a Wikipedia article on chaos theory. None of it made any sense, and it wasn't just for lack of sleep. None of it would ever make any sense to me. If only Dorothy was here to help. We could have done this together as a team.

But I was dead to Dorothy now. From time to time, I tried calling her, but there was never any signal. Since my abortive call to her this morning, something in the building had started blocking me. I tried sending messages to her via WhatsApp, both on our private channel and via the group, but I eventually received a private message from Balvinder asking me in no uncertain terms to back off. I tried to explain what had been going on, but he wasn't having any of it. I almost thought of calling Ali, but there was no way I'd get anywhere with her. She had her ups and downs with Dorothy, but with me it was almost 100 per cent downs.

The only person who might have helped was Patrice, Ali's partner. She was a professor of mathematical biology, a subject whose existence I still found it difficult to believe in, but she was a kind, gentle person who I was sure would listen to me with a sympathetic ear. She owed me one, too, as I had helped to save her controversial PhD thesis from the hands of the Fractal Monks. Even if I did accidentally destroy it in the process, by hurling it into the blades of a helicopter.

But I had no way of contacting her.

This was a pity from another point of view, because I remembered now it was Patrice who'd told me that chaos theory was closely aligned with fractals and she might have been able to help me understand both.

Oh, it was hopeless. Tomorrow afternoon I would be picked up at Minsk airport and driven to meet a mafia boss in order to brief him on everything to do with chaos theory, at which point he would realise I was a complete bullshitter and he would order me to be killed. On this basis, I reckoned that I probably had a little over twenty-four hours to live.

It was a shame, because up until the previous night's fiasco, life had been pretty good. I was no longer tied to a terrible job. I was in a relationship that seemed to be going somewhere, even if I didn't have a clue as to the ultimate destination, or indeed any idea of the route either. And, even if I sometimes worried about what we were up to, it was undoubtedly an exciting life.

Now, it was just terrifying.

The only thing I had to hang on to was there was an outside possibility that I might be able to find Sergei and hence get hold of the Vavasor papers and thus redeem myself with Dorothy. Given that Sergei was unlikely to still be alive at this point and, even if he was, equally unlikely to be happy to part with the papers, the chances of redemption were, however, as slim as an undernourished flounder on the wrong side of an altercation with an underwater steamroller.

At around one o'clock, the door opened and Brett's massive figure squeezed into the room, bearing a tray.

'Food,' he said.

I hadn't realised he could talk. This was an exciting new development. Maybe I could engage him in conversation and somehow outwit him.

'Cheers, Brett,' I said, as he put the tray down on the table. 'It is Brett, isn't it?'

Brett grunted.

'Do you like it here, Brett?' I said.

He grunted again. This wasn't going quite as well as I'd hoped. Maybe if I steered clear of questions with yes/no answers I'd do better.

'How long have you worked for Matheson?' I said.

Brett counted on his fingers. He held up three chunky digits.

'Right,' I said. 'Right.' Then I remembered being sent on an interview course one time and being told that open questions were always better for establishing a dialogue than closed ones. I racked my brains.

'So what's the best thing about working for her?' I said.

'Huh?' Brett frowned at me and I could almost see his brain trying to figure out what to say. But after a minute or so, he gave up and shrugged. 'Eat food,' he said, pointing to the tray. 'OK?'

'Yeah, fine,' I said, as he turned and left the room. This was disappointing. We probably didn't have much in common beyond our shared employer, but I still felt we had so much more to say to each other. However, I didn't feel as if I'd made any headway at all towards building a rapport with the man.

The food turned out to be a gloopy chicken stew with rice and some vegetables that had been boiled into extinction, but it filled me up. I hadn't realised quite how hungry I'd become. The last twenty-four hours had been a confusing time for my body as well as my brain.

After half an hour or so, Brett reappeared to take the tray away. I wondered if my attempts at befriending had had some effect after all, because he was clearly making an attempt to smile as he approached me, even if the outcome was ambiguous and ever so slightly sinister.

'Good?' he said, waving a pudgy hand at the empty plate.

'Very nice,' I said. 'Is the food here always this good?'

He grunted by way of response, then picked the tray up and left the room. It would clearly be some time before we got round to any sort of critical discussion of the finer aspects of the local cuisine.

A couple of hours or so later, Matheson dropped in.

'Darling!' she said. 'I do hope they are treating you well.'

'Lunch could have been better,' I said.

'You have no idea.' She rolled her eyes in a conspiratorial manner. 'The housekeeper does all the cooking herself and she has a somewhat limited range. I've been trying to teach her, but she just won't listen.'

'Why are you doing this to me?'

'Doing what?'

'Wrecking my life.'

'Sometimes, Tom, in order to create something new and wonderful, a few things have to be broken.'

'And am I one of the things that is getting broken?'

'Good lord no, sweetheart. You're the new and wonderful thing that's being created. Can't you see that?'

'No.'

'I'm disappointed in you, Tom.'

'You're mad, aren't you?'

'I prefer visionary. I'm trying to make the world a better place.'

'My world is definitely not a better place right now, Matheson.'

'But it will be, Tom. There's a whole new life opening up for you. You'll see. You may feel as if your life is down in the dumps right now, but one day soon, you'll be in the penthouse of your dreams. Your life will be lit up. Trust me, darling.'

I guffawed at this manifest nonsense.

'Well,' said Matheson. 'If that's how you feel, there's not a lot I can do to change your mind, is there? Anyway, just to let you know, I've got your plane ticket and cash, which I'll give you tomorrow morning on our way to the airport. Their driver will meet you at Minsk International 2 and will be holding up a placard with the name "Mr Wilberforce."'

'Why Wilberforce?'

'Well, they couldn't use your real pretend name, could they? In case someone finds out about Dr Milford being there.'

'Which he won't be.'

'Or anyone pretending to be him, either.'

'Right. So I'm Tom Winscombe, pretending to be Dr Rory Milford, travelling under the name of Wilberforce.'

'That's about it.'

'I'm bound to cock this up, you know. I'm going to die.'

'Stop it, Tom. Look, their driver will be called Yevgenei Smirnov, and you must check that it's really him.'

'What if he's operating under an assumed name?'

'Please, Tom. This isn't helpful. Smirnov will take you to your hotel. You've got a reservation at the Minsk Metropole, which I'm told is comfortable if not quite what it was in its glory days. I've heard a rumour or two about shortcuts taken with the construction, so I wouldn't advise you to outstay your time there.'

'I'm going to die.'

'Tom, I was joking. You'll be fine. Once you've had a night's sleep, Smirnov will come for you after breakfast and take you to your first meeting with the Petrovs.'

'Who are the Petrovs?'

'The family my man is embedded with.'

'Ah. And the main man is?'

'Nikita Petrov. Big fan of greyhound racing. He may want to try you out with some kind of betting strategy, so have a think about what you could come up with that looks convincing.'

'I literally have no idea what this chaos theory stuff is all about, Matheson.'

'Did you read that article I showed you?'

'Well, yes, but—'

'Well, then. Bleat on about basing your betting strategy around the Lorenz attractor. And then tell him to put fifty roubles each way on the dogs with the names you like the most. I suspect that's what Milford would do, anyway.'

'I'm going to die. Horribly.'

'Tom, will you please stop saying that? It's morbid and boring.'

'It's true, though.'

'We're all going to die at some point, Tom.'

'I'd like it to be a bit further in the future, though.'

'Well, it's up to you to stay safe, darling. Look, you'd best get some rest. As you see, you've got a busy day ahead of you tomorrow.'

Matheson put a sympathetic hand on my shoulder, but I flinched away from it. I wanted none of her bonhomie. All I wanted to do now was get in there, establish contact with Sergei and get out safely. Nothing else mattered.

I lay down on the bed and stared at the ceiling. Was I really going to go through with this? I'd been in a few scrapes recently and somehow I'd survived every single one of them. But I'd had Dorothy by my side then. This time I was heading into the unknown, on a hopeless mission in a foreign country where I couldn't even read the alphabet, let alone speak the language. I really was going to die.

Oh, this was ridiculous. Maybe I should abandon the whole idea, forget about trying to redeem myself by retrieving the Vavasor papers, and somehow try to convince Dorothy that I was telling the truth and that all those WhatsApp messages were indeed planted by Matheson.

In that case, maybe I should try to escape from this place? Perhaps I could contact the authorities and someone would come to rescue me. There were several problems with this. First of all, I had no guarantee that the authorities would be friendly to me, because there was every chance that Matheson had connections there from her previous life. Secondly, I had no signal on my phone and I suspected that even if I could get through to the right person by email or whatever, they wouldn't take any notice of me.

Finally – and this was the clincher – I had no idea where I was. The internet here seemed to be anonymised via some kind of VPN arrangement, so that my location on Google Maps was constantly changing. At one point, my position seemed to be a few miles south of Svalbard, while an hour or two later, I'd apparently moved to downtown Dar es Salaam. A glance out of the window earlier had revealed no notable landmarks or even street signs. The only remotely distinctive object in sight was a flaccid weeping willow in the front garden of the house opposite.

No, if I wanted to escape, I was well and truly on my own.

But surely Brett wouldn't be squatting outside my room the entire night? I got up and went over to the door. I pressed my ear to the keyhole and listened hard. There was undoubtedly someone or something breathing laboriously very close to the other side. It was either the sound of someone who was fast asleep or the sound of someone whose massive body required a large intake of oxygen. I couldn't take the risk that it was the latter.

I turned round and surveyed the room. Maybe I could exit via the window? Tie my sheets together and climb down? I stood up and walked over to the single window on the opposite side of the room. There was a sturdy lock on the catch, and even if I'd managed to lever it open somehow, there were those solid steel bars on the outside. Bit of a fire risk, now I came to think about it, but I guessed that Matheson and her crew weren't so bothered about that kind of thing. I was, in the end, largely expendable.

No, I didn't have a choice about this. I was going to have to go through with it, whatever happened.

By the time Brett turned up with my evening meal, I was too exhausted from lack of sleep to attempt to engage him in any further conversation. I wasn't even sure what it was that was on the plate in front of me, and I only managed to consume half of it before my body finally convinced me that what it really, really wanted to do right now was switch off completely. So I crawled into bed and slept like a log until I was roused by Brett bringing in the breakfast tray.

'Morning, Brett,' I said, in a renewed attempt to capitalise on our burgeoning friendship. He merely grunted by way of reply, put the tray on the table and walked out. Perhaps he wasn't a morning person. I could relate to that.

I got up and stared at the Weetabix and toast on the table and decided this was a rather poor effort for the last breakfast of a condemned man. So maybe this was fate trying to tell me that I wasn't condemned after all. Or maybe it was just that they didn't make much of an effort with breakfast around here. I drew back the curtains to let the weak autumn sun shine in, then sat down to eat.

After a while, Brett reappeared and took the tray away again.

'I'm off to Belarus today,' I said. Brett furrowed his brow at this.

'Russ?' he said.

'Um, yes. Belarus.'

'Russ.' He gave me a thumbs up and left the room. Another year or two here and we'd be debating the finer points of Metaphysics and Epistemology, I was sure.

After a while, Matheson dropped in pulling a bag on wheels behind her.

'Clothes, darling,' she said, gesturing towards the bag. 'Should fit you. Borrowed them from one of my boys who's around your size.'

I unzipped the bag and peered inside.

'Great,' I said. 'Including the underwear?'

'Especially the underwear, Tom. Unless you fancy going commando.'

I briefly considered the shame of my corpse turning up in a morgue somewhere wearing dirty underwear. Someone else's dirty underwear, too.

'I might just do that,' I said.

'Gosh,' said Matheson. 'How exciting!' She bent down and unzipped a compartment in the front of the bag. 'Now,' she said, pulling out a cardboard folder. 'Here's your passport, Dr Milford. We've tweaked the photo so that it looks a bit more like you. Shouldn't be a problem getting through. Here's your flight details. Open return.'

'Glad to see you're expecting me to come back,' I said.

'Tom, it's Dr Milford's booking. Don't take that as any guarantee. But I'm sure you'll be just fine.'

'Thanks. That makes me feel so much better.' The mention of Milford's name reminded me of something important. 'Who am I supposed to be pretending to be when they pick me up?' I said. I'd forgotten already.

'Wilberforce,' said Matheson, patiently.

'And the name of the person who's picking me up?' I really hadn't taken any of it in at all.

'Tom, were you actually listening to a word I said yesterday afternoon?'

'Not much, to be honest.'

'I didn't think so. Your eyes had a distinct glaze to them.' She sighed. 'It's Yevgenei Smirnov. And for heaven's sake, make sure it's him. Whatever you do, don't drive off with the wrong person.'

Matheson went back to the folder and pulled out a stack of banknotes. 'And some roubles for you. Make sure you keep receipts for everything. The accounts people round here are absolutely psychotic.' She stood up and brushed her hands together. Then she looked at her watch. 'Right,' she said. 'Five minutes, then we're off to the airport.'

'Five?!' I said, aghast. I wanted more time to prepare myself. Like a year or two.

'No point in hanging around, darling,' said Matheson, as she left the room.

I sat on the bed, paralysed, for a good couple of minutes before I managed to snap into action. I washed myself and picked out some clean underwear from the selection available and ruffled my hair into something close to the style that Dr Milford would go for if he'd been forced to have a trim. I had just applied the final tweaks when Matheson reappeared, followed by Brett.

'Phone, please,' she said, holding out her hand.

'Why?' I said.

'Can't have you phoning for help while we're on the way there, can we?'

'As if I'd do something like that,' I said, handing it over to her. Of course, that exact thought had occurred to me, but I'd also assumed that she'd have thought about it too, so it wasn't a great surprise.

'And if you don't mind, we're going to blindfold you. Brett!'

'I do mind, actually,' I said.

'I thought you might, but unfortunately, there's nothing you can do about it.'

Meanwhile, Brett had placed a strip of cloth in front of my eyes and was busy tying it, none too gently, at the back of my head.

'Right?' said Matheson. 'Are we ready for the big adventure?'

I said nothing. Brett grabbed me by the arm and steered me down the stairs and into the back of a car. There was no turning back now.

Chapter 7

The flight was, thankfully, uneventful, and we landed at Minsk International 2 Airport right on schedule. I collected my bag and found my way out of the terminal, where there was already someone holding up a sign with the legend 'WELCOME MR WILBERFORCE' on it. Unexpectedly, the person holding it up was female. She was tall, with long black hair and unnecessary sunglasses, dressed in an understated dark grey suit. For a moment, I wondered if I'd misgendered the name Yevgenei, but I dismissed the thought as soon as it had arrived.

No, this wasn't Yevgenei.

Maybe there was another Mr Wilberforce on this flight. I scanned the line of drivers awaiting their clients, but there was only one sign with that name among them. Ah well, maybe Yevgenei couldn't make it. I walked up to the woman.

'Wilberforce,' I said, holding out my hand. The woman didn't take it, but gave the slightest of nods and turned away, walking towards the exit. I followed her as she strode away in the direction of the car park. Every now and then she turned and looked back at me, as if to check that I hadn't done a runner.

The car park was a large open space a short walk away from the main concourse and my driver had parked right at the far end of it. The car was a silver Toyota Corolla from, I estimated, a few years back, but as we drew close, I noticed that the driver's seat was already occupied.

'Ah,' I said. 'Yevgenei?'

The woman smiled back at me. 'Yevgenei,' she confirmed.

I threw my bag in the back seat and climbed in alongside. The woman closed my door but didn't go round to the passenger side as I'd expected. Instead, she opened the driver's side door, grabbed Yevgenei by the scruff of the neck and hauled him out onto the ground. Unexpectedly, he didn't put up any resistance, although the reason for this soon became obvious. Yevgenei was dead and had been for a while. There was a single bullet hole in his temple and a look of surprise on his face.

Clearly, something had gone seriously wrong with my arrangements. I had to get out of there. But when I pulled at the handle to let me out, nothing happened. The child safety catch must be on. I tried the door on the other side and still nothing happened. Shit. I was trapped.

Meanwhile, my new driver had enveloped the front seat in a plastic bag that she'd clearly brought along on the off-chance that she might have to cover up some bloodstains and was now ensconced in the driver's seat. She turned the key in the ignition and moved forward. Then she executed a five-point turn out of the parking space, in the course of which she drove over Yevgenei's body at least twice, judging from the sickening bumps underneath the wheels.

As we roared away from the airport, I slumped back in my seat, wondering what to do now. I had expected things to go horribly wrong at some point during the operation, but I hadn't anticipated it happening quite so early on.

'Who are you?' I said.

The woman said nothing.

'Are we going to meet the Petrovs?' I said.

The woman seemed to find this very funny, so I gathered that we were not in fact going to meet the Petrovs. I surreptitiously slipped my phone out of my pocket and started to text Matheson. But before I could send anything, the driver turned round in her seat, pointing a gun at me.

'Phone,' she said, indicating the front passenger seat with her gun. I wasn't sure which was more terrifying – the fact that she was pointing a gun at me or the fact that we were now wobbling from side to side on a busy motorway with no one currently in control of the vehicle. A loud horn caused her to turn round and adjust direction just in time to avoid colliding with a large articulated lorry. She fired off a torrent of abuse in the direction of the driver, then turned back to me, still with the gun in her hand.

'Phone,' she said again.

I didn't have any choice, so I threw it into the front seat. She opened the window, picked up the phone and hurled it out.

'Hey!' I shouted. The woman just laughed again.

Shit. After less than an hour on Belarusian soil, not only had I been kidnapped, but I was now without any means of communication with the outside world. I was also acutely aware that things would very likely get worse as the day went on.

'Where are we going?' I said.

The woman said something in Belarusian that took a long time to say and accompanied this by waving vaguely in front of her. This was not helpful.

'Minsk?' I said.

'No Minsk,' she said.

Things were definitely not getting any better.

After half an hour of driving, we pulled off the main carriageway and followed a minor road through a village into a small wooded area. There were picnic tables and one other car, a silver Jeep with several years' worth of mud on the tyres, but I couldn't see anyone else around. It was too cold for anyone to be eating in the open today. The driver stopped the car and motioned to me with the gun to get out. I grabbed my bag and did as I was told.

Then she went over to the Jeep, opened up the back and took out a can of petrol. She proceed to pour this over the Toyota, shaking the can to make sure every last drop was used. Then she stepped back, took out a box of matches from her jacket pocket, lit one and flicked it neatly towards the car. There was a 'woompf' sound as the petrol caught fire, and I instinctively stepped back.

The woman took hold of my arm and steered me towards the Jeep, indicating that I should get in the front passenger seat. I stowed my bag in the back and got in. As we drove away, the petrol tank in the Toyota exploded and the woman turned to me with a wild grin on her face. She was clearly someone who enjoyed a good explosion.

'I am Anya,' she said.

'Hi Anya,' I said. 'Can you tell me where we are going now?'

Once again, the reply that Anya gave was long and detailed but entirely incomprehensible. Not that it would have helped, because the only place I knew of in Belarus was Minsk. The position of the setting sun combined with the time of day suggested that we were heading vaguely south, but that wasn't a great help either.

Anya seemed in no hurry to get back to the motorway, preferring to stick to minor roads. There wasn't a lot of traffic, however, so wherever it was we were going, we were making good time. After an hour or so, however, I became aware that what was almost certainly a police car had come up behind us and was flashing its lights to get us to pull over.

I glanced across at Anya, who seemed to be debating with herself as to whether to try to outrun them or to comply with the order. At first, she decided to go with the former option, pressing hard on the gas and accelerating away from the cops. I could understand her logic. I was unfamiliar with the Belarusian criminal code but given that she was potentially in the frame for murder, theft and criminal damage already today, there wasn't much to lose by adding speeding and evading arrest to the mix.

It was, however, also somewhat terrifying, and I was distinctly relieved when the approaching siren behind us announced that the police had decided to take this seriously and give chase. Anya muttered something under her breath which I took to be some kind of Belarusian imprecation and slowed the car down to a halt. She took her gun and waved it at me to indicate that I should stay where I was, then put it away and got out of the car. The police car pulled in behind us and a uniformed man got out and slowly walked towards her.

For a moment my spirits rose as I anticipated Anya's imminent arrest and the end to this whole ghastly saga, but as I watched the encounter between the two, I noticed there was something wrong in the body language. Surely not. I had to warn him. But before I had time to undo my safety belt and get out of the front seat, Anya had whipped out her gun, loosed off two crisp shots, one to the head and one to the heart, and put the gun away again. Then she got back in the car and drove off again.

So today's tally was now two murders (one cop, one civilian), criminal damage, speeding and evading arrest. And it still wasn't tea time. I tried to speak, but my throat was dry and in any case my mouth was hanging open and wasn't in any state to form the necessary shapes required for adult conversation. Fortunately, Anya didn't seem to be in the mood for discussing the afternoon's events either, so we continued on our way in silence.

The next couple of hours passed without incident, apart from the fact that my stomach began to rumble very loudly at frequent intervals. Whenever this happened, Anya chortled and muttered something to herself. Eventually, she pulled in at a small garage that was still open at this time of the evening. I was immediately on the alert for opportunities. I was fully aware that Anya wasn't too fussed about collateral damage, but surely petrol stations had CCTV didn't they? It would be quite hard for her to get away with gunning me down here, and I reckoned I stood a decent chance of making my escape.

I watched carefully as she got out of the Jeep and began filling up the tank, waiting for the moment when she had to go into the shop to pay. When she had finished, she opened the door on her side again and went to grab her purse. I made to get ready to leave the car, but after a second's thought, Anya decided she was having none of it. Instead, she reached over to the glove compartment, took out a pair of handcuffs and clamped me to the steering wheel.

'You could trust me, you know,' I said.

She shook her head. Even if she didn't understand a word I was saying, she could tell I was trying to pull a fast one. She held up a warning finger, slammed the door shut and headed over to the shop. She returned with a couple of stale sandwiches and a packet of crisps. She unlocked me from the steering wheel, passed me one of the sandwiches that appeared to contain beef and some kind of pickled vegetable and drove off once more.

Even though I was distinctly ambivalent about the taste of the food, I devoured it with considerably more enthusiasm than it warranted, along with most of the bag of crisps. I offered Anya one or two, just to show I had no hard feelings, but she wasn't bothered. Apparently killing two men in cold blood didn't give you an appetite after all. Who knew?

I must have dozed off for a while, because the next thing I knew was that we were travelling through a dark wooded area. Even in the poor light, I could feel there was an eerie quiet about the place, as if nature had suffered some kind of trauma in its recent past. More prosaically, I also realised there was something I needed to do quite urgently.

'I could do with a pee,' I said.

Anya turned and frowned at me.

'A pee,' I repeated. 'You know.' I tried to amplify this with a mime, which wasn't altogether successful, judging from the look on her face. I refined my performance slightly and this time, after a brief expression of intense disgust, she realised what I was on about. For a few seconds she seemed to weigh up the pros and cons of confining me to my seat with a bursting bladder before bringing the Jeep to a halt at the side of the road.

'Out,' she said, pointing the gun at me while opening the door on her side.

I got out of the car and the two of us walked a short way into the woods, with Anya holding her gun in one hand and a powerful torch in the other.

'OK, you pee now,' she said.

I looked at her, holding my hand against my eyes to avoid being blinded by the light of the torch. 'You could at least turn round,' I said. I made a circular motion with my fingers. Anya continued to stare in my direction, so I turned back and attempted to do the business.

I was absolutely desperate by now, but of course in the circumstances, in a dark forest with a mad female psycho pointing a gun at me, everything decided to clench up and I was suddenly a seventy-year-old man with a dodgy prostate.

'You OK?' said Anya. 'We go now.'

'Hang on!' I said. 'Nothing's happening yet!'

Come on, Winscombe. Think of running water. Niagara Falls. Angel Falls. Victoria Falls. Surfer's Paradise. White water rafting. Just a fucking tap turning on, for Christ's sake Winscombe. Just anything. Oh, thank god. Here it comes. Here it comes.

'It's OK!' I shouted, 'I'm OK! It's all good!'

I shook off the last few drips and turned back towards the car. Right. I was ready for the rest of the journey now. At the next junction, the car's headlights lit up a road sign telling us that somewhere was seventy kilometres ahead of us.

'Where are we heading?' I said, trying and failing to make sense of the Cyrillic characters.

'Pripyat,' said Anya, and some ancient connection began to surface in my brain.

Pripyat. Where had I heard that name before? It took me a while, but I eventually remembered. Rufus Fairbanks. That was it. He'd been involved in some kind of scheme involving shale gas in the Pripyat basin, along with some shady friends of his. But that wasn't the only connection my brain was making. There was something else, if only I could reach it.

But in the end, I didn't need to reach it myself, because another road sign did all the work for me. As well as Pripyat, it was now showing the way to another, much more famous place.

'Chernobyl,' said Anya, with a grin.

After half an hour or so, Anya turned off the main road onto a dirt track that led through the woods. She drove on for a few miles and then we came across a wire fence with barbed wire curled along the top. There was a gate in front of us, with a rusty sign attached to it, along with a padlock the size of a man's hand. I still hadn't got used to the Cyrillic script, but the tone of it suggested a translation along the lines of 'For Pity's Sake, Keep the Fuck Out: You Seriously Do Not Want To Be Anywhere As Radioactive As This, Trust Me On This One.'

'Are we in Ukraine now?' I said. I was pretty sure that was where Chernobyl was, but we hadn't come across any frontier posts.

'No,' said Anya. 'Is Belarus.'

Well, I guess radioactivity was under no obligation to respect borders.

Anya rummaged in the glove compartment and took out a key. Then she got out of the car, undid the padlock and opened the gate. Once we were through, she closed and secured the gate again and we drove on in silence. After another hour or so of driving through the woods we came to a kind of encampment encircled by more barbed wire and lit by floodlights. The fence was broken by another gate and as we approached, a man in military gear carrying a rifle walked up to the other side.

Anya stopped the Jeep and got out, walking over to our side of the gate. The man on the other side shone a torch in her face, nodded and proceeded to open the gate. There was a brief discussion between him and Anya, followed by a long and passionate embrace. Then they separated and he waved us through, following us on foot.

Once Anya had parked, the man came over to my side of the Jeep and opened the door. He motioned for me to get out with his rifle and I complied. I went to get my bag from the back, but he shook his head and took me firmly by the arm.

'Come,' he said, directing me towards a concrete building ahead of us.

'My bag,' I said. 'I need my bag.'

The man just shook his head and continued to steer me towards the building, with Anya following close behind.

As we got closer, a young woman emerged and walked towards us, saluting both Anya and the man with me. She was also wearing army fatigues. I watched as she went over to the Jeep and took out my bag. So maybe it was just a case of valet service, and I briefly felt bad that I didn't have any loose change to give her as a tip.

Considering the time of night, the main building was quite a hive of activity. Men and women, mostly in army uniform, were walking purposefully from one room to another, carrying important-looking sheaves of paper. I was shown into a small office two doors down the main corridor and directed to sit in a chair opposite the desk. The man and Anya took up positions behind me, one on each side. The room was lit by a single naked bulb hanging from the middle of the ceiling. The walls were all painted white and the floor was bare concrete. I guessed that if you were putting together a facility in the middle of the Chernobyl fallout zone, interior design would perhaps be low on your list of priorities.

The other side of the desk was occupied by a middle-aged woman, who was scrutinising me very closely, as if she didn't quite believe what she was seeing. I waited for her to open the conversation, but she didn't seem in any great hurry. From time to time she glanced down at the contents of a folder on the desk in front of her and nodded.

Then the young woman came in with my bag.

'Ah,' I said, getting up from my chair. 'Thank you!' But before I could stand up fully, Anya grabbed my arm and hauled me back down again. So it wasn't valet service after all.

The young woman placed the bag on the ground next to the desk and proceeded to unzip it. Then she went through the contents in turn, subjecting them to close scrutiny. As she held up each item, she and the woman behind the desk conducted an animated discussion in Belarusian, presumably to establish the merits or otherwise of the object in question. This became more than a little embarrassing when it came to the underwear.

'They're not mine,' I said. 'I borrowed them.'

Both women looked at me and shared a look that suggested I was some kind of lunatic, before resuming their conversation. On reflection, it probably wasn't a helpful intervention on my part.

In the end, the only thing they found fault with was my passport. The woman behind the desk took this, waving it at me accompanied by a long string of Belarusian, which I took to mean something like, 'Well, matey, it's unlikely you're going to get out of here alive, so you won't be needing this any more, will you?' Then she put it away in her desk drawer.

Under any other circumstances, the loss of my passport would have been sufficient to bring on a hyperbolic anxiety attack, but quite honestly this was the least of my worries right now. I'd been kidnapped on my arrival in this country by a psychopath, witnessed one murder and the after effects of another and been driven for miles away from my original destination, only to be delivered into the care of what appeared to be some kind of militia hiding out in a radiation zone. It must be near midnight, surely? There couldn't be any more surprises lurking in the darkness, waiting to pounce on me today, could there?

I was wrong.

The woman behind the desk indicated that tonight's meeting was over and that it was time for me to go. I stood up and picked up my bag. As I turned to leave the room, it struck me that there was a very important question I hadn't asked yet.

'Who are you?' I said.

'My name is Yuliya Gretzky,' she said.

Chapter 8

I woke up next day with a headache. Logic told me it was probably because I hadn't drunk enough the previous day, but emotion told me that it was caused by my brain struggling to cope with the seemingly endless tsunami of difficult information that it was having to process.

I still wasn't entirely sure of my status here. I was certainly being held against my will, because if my will had anything to do with it, I would have been waking up in a comfy room at the Minsk Metropole rather than the featureless box that I had just spent the night in. But however ugly this room was, it wasn't a prison cell. The key to the door was on my side of the lock.

However, this would all change when the inevitable time came when they discovered that I was not the mathematician they thought they'd kidnapped and that whatever problem they had was going to remain unsolved for the foreseeable future. It was also likely to change if they ever realised that I was the one who delivered the bomb that blew up a significant proportion of their family. I guess it was possible that Yuliya Gretzky might be grateful to me for providing the means for her to ascend to the top job, but it would have been optimistic of me to rely on her goodwill if the truth ever came out.

I noticed that a set of regulation army fatigues had been laid out for me to wear. I decided that for the time being at least I should make an attempt to fit in, so I got dressed in them and sat on my bed, taking a few deep breaths and trying to prepare myself for the day ahead. I had to be ready to pass myself off as a top-flight mathematician, so I needed to start thinking like one. Also, if I was going to try to escape from this place at some point, it would be best if I were to do it as soon as possible, given the likely changes in my status once the truth about me began to come out. So I also needed to plan how I was going to achieve that.

Finally, I needed to work out what the hell was going on here. I could hear some noise coming from outside, so I went to the tiny window in the opposite wall to the door and looked out. There was a kind of parade ground out there and a motley bunch of men and women were engaged in some kind of early morning drill. This involved a lot of enthusiastic shouting and stamping, although there was a fair bit of work still to be done on the coordination front.

Was this some kind of militia training camp? Were the Gretzkys forming their own private army? Who were they planning to take on? The Petrov family, perhaps? That would certainly explain why they'd intercepted me at the airport, if they had spies in the enemy camp. But I wasn't going to get any answers to these questions by staying where I was, and besides I needed some breakfast.

I opened the door of the room and looked out into the corridor. There was no one there. I sniffed the air and caught a whiff of something edible, so I followed my nose down the corridor towards the back of the building and out into the yard outside. I pulled the collar of the army jacket closed against the cold morning, watching my breath steam out into the clear air. Or perhaps make that clear, yet presumably scarily radioactive air. I reminded myself that it would be a good idea to get out of this area as soon as possible.

A series of trestle tables had been set up in the yard and orderlies were bustling around the two on the far left-hand side, positioning a couple of vast metal urns and several trays of a dubious porridge-like substance, which looked as if it would serve quite well as a substitute filler for some of the cracks that were apparent in the exterior of the main building. There was currently no one else around, as the morning drill was still in progress, judging by the noise coming from the adjacent area, so I grabbed a mug and plate and helped myself.

I sat down at one of the tables near the back and took a spoonful. It tasted better than it looked, and whatever it was, it was warm and filling. I devoured the lot and wondered whether or not I would be allowed to go back for more. I decided that it would be best not to draw too much attention to myself, so I remained where I was and sipped at my coffee, which was warm, sweet and milky.

Beyond the yard where I was sitting were a large array of small huts with large, naïf pictures of flowers, animals and children painted on the sides. This must be the main accommodation area where the Gretzky army hung out. I suddenly realised that this place must have been some kind of children's outward-bound camp back in the days before the plant blew up.

It also struck me that I was evidently being treated like some kind of honoured guest at the moment, and I felt sorry for whoever had been thrown out into the children's dormitories to make way for me. Still, they'd be back soon enough once I'd been found out. Or once I'd escaped.

At some point, I would have to explore the perimeter of the area more fully to examine it for weaknesses. If it was just wire holding us all in, there must be some bolt cutters somewhere in the camp. All I needed to do was find them, sneak out at night and snip myself free. There was, of course, the small matter of finding my way back to civilisation through the radioactive woodland, but that was a minor consideration compared to what was likely to happen to me if and when I was found out.

My thoughts were interrupted by the arrival of the rest of the company. There was a lot of jostling around as they collected their porridge and found somewhere to sit.

'May I sit with you, Dr Milford?' The voice belonged to a fresh-faced young man who looked all of eighteen. For a moment I was confused, as the question – delivered in perfect, barely accented English – was clearly directed at me. Then I remembered who I was supposed to be impersonating.

'Oh, yes, sure, do take a seat,' I said, waving my hands around in a manner that I hoped suited that of an eccentric academic.

The young man put his food down opposite me on the table and sat down.

'Alexei,' he said, offering a hand. I shook it. He had a firm grip and when he looked at me there was fervour in his eyes.

'Pleased to meet you,' I said.

'The pleasure is all mine, sir. It is an honour to meet such a distinguished academic.'

I wasn't sure how to respond to this without sounding like a complete idiot, but fortunately Alexei hadn't finished.

'It was my idea to bring you here,' he said, tucking into his porridge. 'My apologies for the inconvenience.'

'Right,' I said.

'I hope you are being looked after?'

'Um, yes. I suppose I am.'

'Good.' Alexei looked at me and put his head on one side. 'You look different from the picture on your website,' he added.

Well, there was a very good reason for that, but I said nothing.

'You look younger,' he said.

'Ah well,' I said. 'The camera always adds a few years, doesn't it?'

'Ha yes. I have heard this is true. But you must have been very young when you worked with the Vavasor twins, no?'

Holy shit. There were obviously some aspects of Dr Rory Milford's career that Matheson had omitted to tell me about, including the one where he'd worked with the sodding Vavasors. The Vavasors who had both died when I was still at school.

'Well, I—' I began.

'But then, I guess you were something of a prodigy,' said Alexei, taking another large mouthful. 'Like me.'

Oh Christ. I'd only just finished breakfast and I was about to start swapping equations with a real-life mathematical genius. My life was going to be over before morning coffee.

'I have always been fascinated by the Vavasors,' said Alexei. 'Such brilliant careers and such tragic deaths. And no one knows why. Unless you do?' He laughed, as if it was a completely preposterous idea.

Oddly enough, I was one of the very few people who actually did know the truth behind the deaths of Archimedes and Pythagoras Vavasor. But this probably wasn't the time to mention it, as it would lead to all manner of questions as to exactly how I came to know, and my brain didn't feel up to improvising an extended lie this early in the day.

I shook my head, with a sad smile. 'It was a terrible shame,' I said. 'I was so fond of them both.' Well, there it was: my first genuine lie of the day, rather than the assumptive silences I'd previously committed. I was not remotely fond of the Vavasors. The bastards had blighted my life, and I had serious doubts that they were indeed good people.

'I met them when I was very young,' said Alexei. 'But I do not remember much. Tell me about them.'

'Who?'

'The Vavasors.'

Oh Christ. I racked my brains to find anything useful that I might have learnt about them over the previous few months. Right on cue, my brains went blank and put up the anti-racking barriers. What had Patrice said about them? Ali's partner had been one of their last PhD students and I was sure she'd told me a few things, but nothing useful was forthcoming.

'They had a sweet little cat,' I said eventually. 'Called μ.' That wasn't wholly true, either. Of all the adjectives that could be applied to μ, 'sweet' most definitely wasn't one of them. 'Feisty' or 'vicious' perhaps came close, although that didn't encapsulate the full extent of her personality.

'Ah, I remember μ! Whatever happened to her?'

What happened to μ was that I had retrieved her from the home of the late Mrs Standage, the Vavasors' ex-housekeeper, following her murder at the hands of Rufus Fairbanks, during the killing spree that he initiated when it looked as if the truth about Archie and Pye might come out and that he would be publicly associated with the Gretzkys. μ was now in the care of my father at his mobile home, where she dominated the entire household, up to and including Wally, my father's large and odoriferous dog.

'I have no idea,' I said. 'She's probably no longer with us.' I was beginning to feel more relaxed about the lying thing. Maybe I might get through this after all.

'Ah, that is sad, too. But you are probably right. I think she went to live with Mrs Standage, yes? Did you meet her?'

Oh god. Alexei was a full-blown Vavasorologist.

'I met her once,' I said. 'But she was very quiet.' This wasn't entirely false, although the reason she was quiet on the one and only occasion that I encountered her was that she was in the middle of a failed attempt at CPR by a couple of paramedics.

'But what were Archie and Pye really like?' said Alexei, who clearly wasn't in the mood for letting me off the hook quite yet.

I took a deep breath. 'Well, they were two very eccentric men,' I said, 'as you might imagine.' I paused, waiting for inspiration to strike. Alexei gave me an encouraging smile. Come on, Tom. Just let it flow. Ah, hang on. 'It was sometimes as if they had their own secret language.' Yes, that was good. Twins. Secret language. All good stuff.

'I can imagine this,' said Alexei. Yes, I could too, Alexei.

'And when they were working together,' I said, 'the atmosphere was very intense and it was very hard for an outsider to break in.'

'I can imagine this also. How did you tell them apart?'

Bloody hell, that was a curveball. 'Um… one of them had a scar. Just above… oh, where was it?'

'I remember this now,' said Alexei, and I inwardly high-fived myself. 'It was Archie, was it not? On his left cheek.' He touched the place and I mirrored the action.

'Yes,' I said. 'He fell off his bicycle when he was little.' Ooh, nice touch, Tom.

'Really?' said Alexei. 'I thought it happened in a fight at school.'

Shit. Reverse. Reverse.

'Ah, yes,' I said, improvising wildly. 'They were fighting about his bicycle. That was it.' Terrible, Tom. Just terrible.

'Ha, I can just imagine this,' said Alexei. 'Kids can be so cruel.'

'Oh, they can,' I said, adopting a serious tone. 'They can.'

Fortunately, at this point we were interrupted by the arrival of Anya.

'May I join you?' she said.

I stared at her in disbelief. 'You speak English?'

'Why yes. Is this a problem for you?'

'But yesterday you didn't.'

She shrugged. 'Yesterday was difficult day,' she said, taking her place next to me at the table. You're not kidding, I thought to myself. You killed two people yesterday, Anya. That's a difficult day for everyone concerned.

'Right,' I said. 'Right.' This was all doing nothing for my headache.

'But today will be a good day,' she said. 'Did you sleep well?'

'Sorry?'

'Did you sleep well?' She tucked enthusiastically into her breakfast, while Alexei muttered something about having to be somewhere else and hoping to see me later. He picked up his tray and left Anya and me together.

'You threw my phone away,' I said. This was all getting a bit too chummy.

'I am sorry,' said Anya. 'Security. This place very secret.' She gestured vaguely around her.

'You could have just switched it off.'

She shook her head. 'Not safe enough.'

'They're expensive to replace.'

'Is the rules.' She took another large mouthful of porridge. 'He's a good boy,' she said, gesturing with her spoon at the departing figure of Alexei. 'His life not easy.'

'Why's that?' I said.

'His father killed not long ago. More coffee?'

'Ah, yes please.' Anya grabbed a jug from the table next to us and poured me out another mugful. I needed all the caffeine I could get this morning, so I took a large swig.

'Yes, he die in England.'

At this point I only just managed to stop myself from choking.

'Are you all right, Dr Milford?' said Anya, giving my back a good thump.

I nodded vigorously and put my hand up while I struggled to recover my composure. Eventually I managed to get my body back under control, and I was able to continue the conversation.

'Where in England did he die?' I said, hoping I was managing to keep my voice level.

'Somewhere in the south, I believe. He very brave man. He try to recover Vavasor papers but they trick him with bomb.'

It wasn't quite like that, Anya, I thought to myself. You missed out the bit about holding Dorothy hostage. Oh god, Dorothy. If only you were here now. You'd know what to do.

'If ever I find any of the bastards that planted the bomb,' said Anya, 'I kill them.' She gave a grim smile and drew a finger across her neck.

'Absolutely,' I said. There was no doubt whatsoever in my mind that she would follow through on the threat.

'So, Dr Milford,' she said. 'Will you work for us?'

'Do I have a choice?' I said.

'Not really.'

'What about Nikita Petrov?'

'What about him?'

'I'm supposed to be working for him.'

'Forget him. The Petrovs are very bad people.'

'They're going to be a bit upset when I don't turn up.'

'This is true. You will have to make sure they don't ever find you, because they will certainly kill you. You'll need to change your name or something.'

I almost burst out laughing at the idea of that.

'So you work for us, yes?' said Anya.

'What are you expecting me to do, though?' I said.

'I'm sure Alexei will tell you soon enough.'

'Alexei?'

'Yes, he is our chief theorist.'

'But he looks about twelve years old.'

'Alexei very clever boy.'

'Right. What if I refuse?'

'We kill you,' she said. 'Sorry, but that's the way it is. You get good funeral and body returned to England free of charge.'

'Well, that's a nice touch.'

'We are not monsters.'

That was the second time I'd heard that in the last few days. I wondered briefly how Matheson was getting on. The thought that she would be extremely unhappy once she found out how things had progressed was the only thing that was giving me any pleasure right now.

'I'm sure you aren't.'

'Also, we pay you well. Here first tranche.' Anya reached into her combat jacket and withdrew a fat wad of Belarusian rouble notes. She handed it over to me, saying, 'second instalment on completion of work.'

For a moment I thought about refusing the money. After all, it was mafia cash and had probably been earned illegally at the expense of others. But then again, it was cash. Untraceable and, judging by the size of the wedge, a lot more than Matheson had given me. It was, after all, very unlikely that I would succeed in doing anything that would actually bring any material benefit to the Gretzkys;

in fact, there was a very high probability of the precise opposite of this happening. On the other hand, there was still every chance that I would die horribly at some point very soon, but if I could somehow survive, I might at least have something to put towards my pension.

'Thank you,' I said, pocketing the notes.

I was committed now. Whether I liked it or not, I was part of the Gretzky family.

Chapter 9

It turned out that Alexei was too busy to talk to me for the next week or so, which meant that I had a few more days than I'd anticipated without being found out and that in turn meant that I had a few more days of existence on earth to enjoy. However, if the truth were told, I would have preferred those few bonus days to have been lived almost anywhere else in the world than in the middle of a radioactive forest somewhere in the middle of Eastern Europe, surrounded by a bunch of trainee psychopaths.

I began to get used to the daily routine. The alarm, in the form of an irritating little electronic arpeggio, repeated several times and then followed by what I assumed was a rousing Belarusian exhortation, would sound at six o'clock in the morning and everyone would make their way to the parade ground for the morning drill. It was made clear to me that while I was staying in the camp, I should join in too. Still, it passed the time and I probably needed to get fitter anyway. Also, the more I threw myself into things, the more chance there was that I might get some kind of insight into what was actually going on here.

The thing was, I really wasn't sure what the hell this place was all about. My previous understanding was that the Gretzky family were just your average bog standard organised crime syndicate, into the usual sort of thing that organised crime syndicates were into. Bit of drug dealing here, bit of prostitution there, bit of extortion over there and so on. And no doubt this still went on back in Minsk. But this was a paramilitary operation. Then again, perhaps this was what they all did. I wasn't, after all, particularly au fait with modern mafia organisational theory, as it hadn't been covered in any of the management courses I'd been on when I was in full-time work doing PR for the meat industry.

What were they up to, though? No one seemed inclined to tell me. Whenever I asked her, Anya would just say something about the necessity to be prepared for every eventuality, and very few of the others spoke any English. One or two looked actively scared when I tried to raise the subject, saying they were frightened to talk in case Novikov found out. This guy Novikov didn't seem to have a first name, which made him appear even scarier. During these brief conversations, I began to wonder how many of the people there were truly willing participants. This would, of course, explain the ragged discipline on display during the morning drill, which got no better as the days wore on.

Once the drill was over, we would have our communal breakfast and then we were divided up into various work teams. Once again, it was made very clear to me that while I was not yet occupied with Alexei, I was expected to help out. This didn't bother me too much, because as well as giving me the opportunity to find out more about the set-up here, I hoped that by making myself part of the furniture, it would help to soften the blow when the inevitable moment arrived and I got rumbled.

After the morning's work, we would all sit down again for lunch, which usually consisted of one of those stews that tends to be graced with the epithet 'hearty'. This basically meant that it was very big on potatoes. Then in the afternoon, we'd be assigned to work parties again, before returning for a final communal drill under the floodlights at the end of the day. Then it was more stew and straight to bed, exhausted.

I tried very hard to get myself allocated to one of the groups that went out of the camp to collect firewood, but unfortunately this activity seemed to be restricted to the older lags. The rest of us were, for very good reason in my case at least, confined to within the perimeter.

On the first day, I found myself in a group sent up to carry out some repairs to the flat roof of the main building. It was only a couple of storeys high, but it did at least afford a decent view of everything around it. This is what I managed to establish. The camp was roughly square in shape, measuring a couple of hundred metres down each side. The main building was close to the middle, but slightly closer to one side, by the only gate. There was a three-metre-high wire fence going all the way round the perimeter, topped with a roll of rusty razor wire. I wasn't sure at first whether this was intended to keep intruders out or to keep everyone else in, although it soon became apparent that both eventualities were in fact catered for.

A range of off-road vehicles were parked at the front of the main building. They were mostly conventional four-wheel drives, but there was the occasional army surplus jeep thrown in for good measure. There was also a large transporter that was used to pick up supplies from a local wholesaler. At one side of the main building there was a long shed with a forbidding lock that one of the others told me was rumoured to house the armoury, although during that first week, I never once saw it open.

On the other side of the building were the kennels, where a pair of wild and possibly rabid dogs lived. These animals seemed to be the result of some terrible and deeply unethical genetic experiment involving Irish wolfhounds, actual wolves and one of the smaller species of bear. For most of the time, these foul beasts from hell were chained up but I also saw them in action once, when one of the kitchen orderlies tried to make a break for it when the gates opened to let the transporter leave the camp. He didn't get very far, and this served as a sharp reminder that these were definitely not nice people and that I should not be lulled into any false sense of security around them.

My work on the second day took me to the latrine pits at the far end of the camp, beyond the dormitory blocks. Inevitably, this area was in a significantly less cultivated state than the rest of the place, but unfortunately, the wire fence was still in pristine condition even here and didn't present any opportunities for an easy exit.

I must have made quite an impression with my contribution to the roof and latrine repairs, because for the remainder of the week I was assigned to help in the kitchens. This was more within my capabilities, as it mostly involved peeling enough potatoes each day to feed a small army. Literally. It also gave me the opportunity to check out the availability of equipment that I could perhaps make use of in my hypothetical escape.

Actually, there wasn't really anything I could make use of. There was absolutely no way I could conceive of using a knife to do any serious damage to anyone and even if I were to try to use one in anger, it was more than likely that I would end up stabbing myself. I couldn't even contemplate using one against either of the ravening canine killing machines.

But by the time I'd finished my stint in the kitchens, the tiniest germ of a plan had seeded itself in the darkest recesses of my brain and I began to think that maybe, just maybe, there might be a way out of this place. However, before I could do anything about it, I was summoned to a meeting with Yuliya Gretzky, Anya and Alexei.

We were in the same room as I'd met Yuliya last time and she was in the exact same position behind her desk, dominating the room. Anya was seated at her right-hand side and Alexei on her left. Alexei had a large stack of papers in front of him. It was like the worst job interview panel I'd ever been faced with and I was half-expecting to be asked some dreadful trick questions about what I considered my strengths and weaknesses.

Instead, however, Yuliya spread her arms wide and launched into a long spiel in Belarusian. The only words I managed to make out were 'Dr Milford' and 'Nikita Petrov'. The former was accompanied by a gesture in my direction and the latter by a mime of spitting on the floor.

I gathered that there was some serious inter-family beef going on here.

After roughly five minutes of this introductory speech, Yuliya came to a standstill and Anya translated. 'Welcome,' she said.

'Welcome?' I said. 'Was that all?'

'It is the gist,' said Anya.

'Right.'

'The gist is important, no?'

'The gist is very important.'

While this discussion was going on, Yuliya adopted a benevolent smile, nodding slightly to each of us in turn. Then she started off again. This time her intervention lasted less than a minute and she ended with the word 'Alexei', waving her hand at him. He responded with a nervous laugh. Anya didn't bother to translate and I assumed that Yuliya had made a hilarious in-joke.

'Dr Milford,' said Alexei. 'I apologise for not being able to speak to you earlier, but I was called away to Minsk.'

'I see,' I said, still not quite believing that this child of half my age was important enough to be 'called away' to anywhere apart from possibly nursery school.

'I have spoken with our strategists in the capital,' he said, 'and, now that we have the estimable Dr Milford on board, they are happy for me to proceed to phase two of our plan.'

'Right,' I said. 'I mean, good. Good.'

'It's more than good, Rory – may I call you Rory? – it is a total vindication of my work to date.'

I nodded to indicate that he was welcome to refer to me as Rory. It was the least I could do, although I hoped he didn't take it as an endorsement of anything else he had said.

'I'm so glad you agree with me,' he said.

Ah.

Right.

'Could you talk me through some of your work, perhaps?' I said. 'Just so I know where you're coming from.'

Alexei's eyes lit up like the pair of twin stars beaming down on planet TOI 1338 b on a hot summer's day. I only knew about TOI 1338 b and the twin suns that it orbited because Dorothy had got very excited when she'd heard about its discovery, and now I was feeling sad again because she wasn't around to appreciate my use of the reference. Was I ever going to see her again? A lot depended on how I managed to navigate my way around the obstacle course that I was about to hit in the next few seconds.

Alexei seemed to hesitate slightly and then thrust the pile of papers in my direction.

'It's all there,' he said.

'Gosh,' I said. 'I don't know what to say.' This wasn't a lie. I literally didn't have a clue what to say.

'You'll recognise a lot of the techniques I have used.'

'I can see that,' I said. I began to leaf through the hefty wodge of paper, which was densely covered in the most utterly terrifying equations I had ever encountered. Wave after wave of nausea swept over me, but I struggled on, trying desperately hard to look as if I understood a single line of it. From time to time, I forced myself to mutter 'Yes', or 'Uh-huh' or, very occasionally, 'Ooh yes, very elegant.' These sounded like the sorts of things that mathematicians might say.

'Do you see what I mean?' said Alexei, when I arrived at the end.

'Very much so,' I said, gasping for air. 'Look, can I take this away with me? I'd like to study it in more detail before commenting.'

Alexei glanced nervously at Yuliya and Anya. Anya gave a slight nod.

'Um, I guess so,' he said with some hesitation. 'I would be happy for you to do so. But be very careful. Is only copy.' He looked as if he was handing over his firstborn into the care of the local ogre.

'Don't worry,' I said. 'I will.'

I walked away from the meeting feeling that I had at least bought myself a little more time. I hadn't actually committed to a particular date or time for reporting back on Alexei's masterwork, so I reckoned I could probably get away with not saying anything for two or three days at least. Possibly more if I was lucky. I asked them for a good supply of pencils and paper, so that I could put up some kind of pretence of working, although if anyone had studied my notes in any detail, they would have realised that I was merely copying out random sections of Alexei's work and annotating them with suitably mathematical-sounding phrases such as 'try flattening the flange function?' or 'use the square!' Sometimes I even convinced myself that I understood what was going on.

However, something happened just after lunch on that very day that threatened to turn everything upside down. I was sitting in my room, scribbling away, when Anya appeared. She had a stern look on her face and she was accompanied by the chap who had met us at the gate on the first night. He seemed to be in charge of security in the camp, so this was a worrying development. What was even more worrying was the fact that he was carrying a clipboard which had the name 'Novikov' on it. So this was the guy they were all afraid of.

'You come with us now please,' said Anya.

'One moment,' I said, gathering my work together and starting to put it away.

'No, you come now,' she said. There was a new urgency to her voice that made me feel nervous.

'Is there something wrong?' I said.

'You come now.'

I stood up and allowed myself to be marched back into Yuliya Gretzky's office. Alexei wasn't there this time, and Anya didn't sit down next to her. Instead, she and Novikov the security guy took up positions behind me, as if to make sure I didn't make a bid to escape.

As usual, Yuliya commenced the meeting with a long speech in Belarusian, punctuated with her trademark sweeping arm gestures and derogatory references to Nikita Petrov. She really had a problem with him. After it was over, Anya spoke.

'There has been a development,' she said.

'Oh?' I said, turning to look at her. There was real anger in her voice as she continued.

'Dr Rory Milford has arrived in Minsk,' she said, 'and he is now with Nikita Petrov.'

Shit.

'But that can't be true,' I said.

'Our intelligence is good. We are sure it is Dr Milford.'

'But I am Dr Milford. You have my passport.' I gestured towards Yuliya's desk drawer. She remained impassive.

'It could be forgery,' said Anya. 'I think you are not Dr Milford.' She came round from behind me and looked hard into my eyes. 'The question is, who are you really?'

Chapter 10

Well, this wasn't good at all. I could sense Novikov cracking his knuckles and limbering up to practise his advanced interrogation techniques on me. The one thing I had to hang on to was that this could not actually be happening. Unless Brett the man mountain had suffered some kind of catastrophic landslip, there was no conceivable way that Dr Rory Milford could have escaped from Helen Matheson and made his way to Minsk. Equally, there was no sensible reason for her to have let him go, unless she had somehow persuaded him to act as an undercover agent for her, and there was no reason whatsoever for him to go along with this.

So there was something else going on. All I had to do was find out what it was.

'I have no idea what you're talking about,' I said. 'My name is Dr Rory Milford and I am here under duress. What possible reason would there be for me to impersonate him?'

'To spy on us?'

'No! I was going to… I was going to *work* for the Petrovs—' the mention of the name caused Yuliya to mime spitting on the floor, '—not you. I hadn't even heard of the Gretzkys until I came here. You have to believe me.'

'So why would anyone want to impersonate you?' said Anya, clearly finding the idea entirely laughable.

'I have no idea,' I said, with a broad shrug to indicate that I found the idea just as ludicrous as she did. She seemed deep in thought for a moment, then she had a long exchange with Yuliya, at the end of which they seemed to arrive at a decision.

'I show you picture,' said Anya.

'You have a picture of him?' I said. This was an unexpected development. A picture of my / Dr Milford's impersonator could potentially be the evidence that cleared my name. It could also just as easily be the thing that condemned me to an unpleasant death.

Anya took out her phone and pulled up her photo stream. Then she shuffled along it until she found the image she wanted.

'Here,' she said, passing it to me. 'Is not good picture. But is him arriving at hotel in Minsk.'

I looked at the picture. There was something curiously familiar about the figure in front of me. I zoomed in a little. Oh no, I thought to myself. You have got to be kidding me. I burst out laughing. Anya gave me a stern look.

'What is funny?' she said.

'Oh, this is hilarious,' I said. 'Believe me it is.'

Anya signalled to the security chief, who moved closer to me.

'I repeat,' she said. 'What is funny?'

'Sorry,' I said, remembering that this room was a considerable distance from anywhere that I might remotely think of as being safe and that the three other occupants would have no compunction in killing me if they felt it to be in their best interests. 'I know that guy.'

'You know him?' said Anya. 'Are you trying to tell us that he is not Dr Milford?'

'Of course he isn't Dr Milford,' I said. 'He doesn't even look like me!' I realised that this was possibly not a convincing argument, but I pressed on nonetheless. 'His name is Benjamin Unsworth and he was last seen working as an archivist and alpaca wrangler for a mad Vavasorologist called Margot Evercreech.'

'Alpaca wrangler?' said Anya. 'What is alpaca wrangler?'

'Someone who looks after alpacas. You know. A bit like llamas.' I attempted to perform a simple mime involving extending my arm over my head to illustrate my point. This didn't help, so I asked for a piece of paper and a pencil. I scribbled out a picture that ended up looking a bit like a goat that had got its head stuck in a drainpipe. All three of them clustered around the picture and turned it this way and that until Yuliya suddenly announced, 'Альпака!'

I gave a vigorous nod and repeated 'Alpaca!' and for a moment, everyone in the room was smiling. Then they all stopped.

'How you know this?' said Anya.

'Um… the Vavasorologists are a close community. I often speak with Margot.'

Actually, I wasn't sure whether or not I was on speaking terms with Margot at the moment. I'd last seen her in the Fractal Monks' monastery, where as far as I could remember we were on the same side, even though she and Benjamin had held me captive for a brief period of time prior to this. However, we had yet to resolve the outstanding issue regarding Dolores and Steven, the aforementioned alpacas, who I'd accidentally stolen from her when I escaped and which were now in the safe custody of my father's dubious acquaintance 'Mad Dog' McFish.

'Why is this person pretending to be Dr Milford?' said Anya. Now this was a very good question indeed, and one which I was struggling to answer. The only thing I could think of was that Helen Matheson had got wind of my failure to turn up at the Petrovs', assumed I was now dead, and had gone through her contact list of available deniable assets to find a suitable substitute. It didn't say a lot for the quality of her list of contacts that the best she could manage was poor old Benjamin.

At this point, she must have got him – in the persona of Dr Milford – to inform the Petrovs that he had missed his flight or got the date wrong or something and then sent him in. Nikita Petrov would have been relieved to find out that, although he had lost a driver, the Gretzky kidnap operation had all been in vain and he would get to work with Dr Milford after all.

'I don't know,' I said, and then a germ of an idea began to form in my brain. Actually, it began to form in my mouth, but fortunately my brain just about caught up with it by the time I got to the end of the story. 'But... I did hear of a rumour to infiltrate the Petrovs by sending in a fake Dr Milford.'

'A fake?'

'Yes, can you believe it? The plan was to kidnap me and hold me captive while they sent in someone in my place.'

The thing I really liked about this lie was that every single word of it was true.

'How do we know this isn't what really happened?' said Anya.

'Because I'm here!' I said.

'But…' Anya paused for a moment, clearly struggling to get everything in order. 'How do we know you weren't kidnapped?'

I stared at her. 'Because I'm here, Anya,' I said, as slowly and as reasonably as I could.

'Yes, but how do we know you're you?'

I spread my arms wide. 'Of course I'm me,' I said.

'What about this Benjamin person?'

'He's definitely not me. I think we can all be agreed on that.'

'What if he is not Benjamin and he is really Dr Milford?'

'What if neither of us is Dr Milford?' I said.

'No!' said Anya, smacking her hand to her head. 'One of you must be. Two fakes not possible.'

'I was joking,' I said.

'Is not funny.'

'Sorry.'

Anya shook her head. I could tell she was on the point of giving up. She said something in Belarusian to Yuliya and Yuliya responded with a long speech, which contained a lot of hand waving and references to Milford, alpacas and Nikita Petrov. The latter name was, as always, accompanied by her miming floorward expectoration. Then she said something else to Novikov, who responded with a grunt. Finally, she waved dismissively in my direction and it seemed that the interrogation was over. I left the office of my own free will and I went back to my room.

The good news was that I had somehow persuaded them to give me the benefit of the doubt on this occasion. The bad news, however, was that I was almost certainly under suspicion now, and that I would have to be doubly careful in my dealings with Alexei. In fact, it would be best if I managed to avoid having any more dealings with Alexei at all, and that meant that I would have to bring my plans forward somewhat. I had to get out of this place as quickly as possible.

I looked at my watch. It was three o'clock in the afternoon. There was a brief period of dead time in the kitchens about now that I could make good use of. I got up and went to the door, opening it up just a crack so that I could peer down the corridor. There was no sign of any activity out there, so I edged out and headed off to the kitchens.

Once I got there, I found a set of whites and togged myself up. Then I marched into the food preparation area as if I owned the place. There was no point in pretending I was somehow invisible.

'Hey, Mister Rory!' called out Ivan, one of the pot washers. 'Where you been? We miss you!'

It was unlikely that Ivan missed me at all. He was a miserable bastard, fully aware that he was stuck in the lowliest position in the kitchen hierarchy and determined to do whatever he could to annoy those who had somehow contrived to rise above him. Fortunately, he had largely left me alone, mainly because he couldn't quite work out whether peeling spuds all day long made my role superior to his or inferior. I wasn't entirely sure myself, but I'd done my best to encourage him in the belief that he and I were in it together.

'Got some other work to do,' I said.

'What work? You Novikov's new fuckbuddy?' Ivan illustrated this with an appropriate mime.

'Er, no,' I said. 'Not this week.'

This seemed to satisfy Ivan, who returned to his washing up without bothering me any further. Apart from Katerina, one of the general orderlies, who was standing by the back door, smoking and staring out into the distance, there was no one else in the kitchen, so no one paid any attention to me as I went over to the freezer and removed half a dozen extremely large T-bone steaks. I'd noticed these a few days back and wondered what special occasion they might have been put aside for. Well, whatever it was, it was going to be a vegetarian event now.

I stuffed them under my chef's jacket, along with the sharpest small knife I could lay my hands on, plus a container of leftover potato stew, and snuck back out of the kitchens and headed off to my room. I passed one or two other people in the corridor but no one paid any real attention to me as I was now as good as invisible to them in my chef's whites. I cached the steaks and the potato stew underneath the bed, hoping that the meat would have defrosted by the time night fell. The knife I hid away in my bag, as a last resort means of defence. Then I went back to the kitchen to change back out of my whites into my civilian clothes.

I passed the next few hours by attempting another assault on the daunting terrain of Alexei's great plan. As far as I could tell, it was all about how small events could trigger much bigger ones, and as I read it, I did vaguely remember something I'd once come across in a book as a kid about a butterfly flapping its wings in Tokyo and causing a hurricane in Florida. It caused a major panic to me at the time, and I remember going to my father and telling him we needed to kill all the butterflies now. He managed to calm me down by telling me that it was only Japanese butterflies we needed to be worried about, although to this day, I'd never been sure as to whether he actually believed this or not.

It was unlikely that any butterfly round these parts was going to cause a hurricane anyway, mainly because there didn't seem to be any. But I could sort of see that an aspiring mafioso family might be keen to look into ways in which it could make use of the theory, although at this stage I wasn't sure whether this was purely to assist them in whatever they were planning with the financial markets, or whether it was a tool for whatever quasi-military operation they were putting together.

Dusk began to fall at around half past six. I changed into my darkest gear and packed away Alexei's papers along with the rest of my stuff. I thought briefly about trying to sneak into Yuliya's office to retrieve my passport, but a moment's consideration told me that it really wasn't worth risking the whole operation for that. It wasn't as if it was real after all.

It was time to go. Everyone, including the dogs, would be having their supper and the supply truck would be arriving back soon. I got the steaks out from under my bed and was relieved to find that they had mostly defrosted. My stomach rumbled and for an instant I imagined myself tucking into one of those T-bones. Or maybe two. It had been some time since I'd last had a meal that didn't consist mostly of potatoes. I shook my head and tucked them into the outside pocket of my jacket. I put the container of potato stew into my bag.

For what I hoped was the last time, I opened my door and checked out the state of the corridor. There was no one there, so I snuck out once more and headed off towards the front door. Then I made a sharp left turn to where Fang and Slobber were finishing off the last of their evening meal. Despite the fact that they had just been fed, as soon as I turned up they began to bark in an aggressive manner and strain with all their might at their chains. The expressions on their faces suggested that they would be very keen indeed to tear me limb from limb as some kind of post-prandial amusement. They were probably kept hungry in order to keep them mean and ready to top up their daily calorific allowance with the odd limb from a passing escapee.

'Ssssh!' I said, quickly reaching into my pocket. 'Here! Look!' The barking stopped immediately and the expression on their faces now changed to one that might be worn by a six-year-old child who has just been informed that, owing to an administrative error, Christmas morning was unexpectedly going to be repeated for the second day running. The realisation that each of them was going to get not one, not two, but three delicious red T-bone steaks only served to intensify the look of sheer pleasure that was spreading over their snouts.

'Make the best of it, lads,' I said, throwing the meat at them from a safe distance. 'Because if you behave the way I'm expecting you to now, you're going to be on hard tack for the rest of the week.' Fortunately, dogs have a fairly short-term attitude towards goal-setting and they didn't seem remotely bothered by this. I left them happily gorging themselves and headed back to lurk in the shadows next to the main building while I waited for the supply truck to arrive. I fervently hoped that it would arrive on time, because if it was late, there was every chance that my absence would be noticed and I would either have to try and hide somewhere or be prepared to come up with a plausible explanation as to why I'd decided to go for a walk at night carrying my bag.

Fortunately, I didn't have to wait for long. After ten minutes or so, an array of headlights lit up the night and there was a flurry of activity at the gate as several of the crew rushed over to help unload the truck. I joined in, keeping my head down, making for the far side of the gate, opposite to where the sentry hung out. I got there just in time before the gate opened to let the truck through. I waited until it was halfway in and then I edged past it out into the wilderness beyond the camp, hoping that the glare of the truck's headlights would stop anyone from noticing me.

I was free.

Chapter 11

Well, this was about as far as my plan had gone, and I now had to come up with some idea as to what to do next. This was actually quite simple to do, because the very first thing I had to do was to run as fast as I could away from the camp and into the darkness so that by the time anyone realised what had happened and turned the searchlights in my direction, I would be well out of range.

This very nearly happened.

Unfortunately, what actually took place was that when I was around a couple of hundred metres away from the camp, I tripped up on a tree root and fell flat on my face, bellowing 'Shit!' at the top of my voice as I did so. This happened to coincide with a moment of silence back at the camp as the truck had now finished moving through and the gate had just clanged shut again.

As I picked myself up, I caught a glimpse of the sentry peering out into the darkness for a brief instant before turning the lights on and catching me in centre stage. Bollocks. I couldn't see any more what was going on, but I heard a lot of shouting and the sound of the gates being opened again. The only thing I could do was keep on running now and hope I could get to some cover before anyone found me. I started off again, zigzagging to try to escape the searchlight as it tracked my path.

Finally, I managed to evade it for long enough to find a decent-sized clump of trees to hide myself behind and I could tell from the frantic back and forth of the lights that they had well and truly lost me. I held my breath and listened out for sounds from the camp. So far no one seemed to have followed me and I very soon realised why as I heard a lot of shouting coming from the area where the kennels were, followed by squealy protests from Fang and Slobber, who had apparently decided that they weren't interested in doing the humans' dirty work for them tonight, thank you very much.

While it was unexpectedly promising that at least part of my plan had come good, the next thing I had to face up to was that there would very likely be a search party heading my way with or without the dogs. The searchlight had temporarily stopped dancing around and was pointing aimlessly some way to my left. Everything to the right of me was in total darkness now, so I decided to risk making a move, although this time I kept to a brisk walk in order to avoid finding myself on the ground again.

I carried on walking for another few minutes until I heard more voices in the distance behind me indicating that they had indeed decided to come after me. As I kept on moving, the voices came closer to me but then veered off to my left. I stopped behind a tree and looked back. Judging by the number of torches being waved around, there were only half a dozen of them. As long as I kept quiet, it seemed unlikely that they would find me now.

I continued on my way. The adrenaline was still flowing hard and it was only just beginning to occur to me that I was now completely lost, deep in an unknown radioactive forest, potentially populated by any number of unexpected local fauna. There were probably wolves and bears and several other animals that justifiably regarded themselves as being higher up the food chain than I was. I had no map, no phone and only enough rations to keep me going for one night and I cursed Benjamin Unsworth for turning up in Minsk. If he hadn't put in an appearance, I would have had more time to plan my escape and I would have made a much better job of it.

However, the awkward truth was that I probably wouldn't have done. I was at heart more of an improviser than a planner, and if I were to be completely honest with myself, this was mostly because of laziness. Planning required so much bloody effort, after all. But it was also because whenever I had actually gone to the bother of making plans, they tended to unravel at the first hint of trouble. If my plans only had the decency to work in the way that they were supposed to do, I'd probably be more inclined to make them.

Now that I was out of the searchlight's glare, my eyes began to adjust to the darkening night. The skies were cloudy and whenever the moon did manage to put in an appearance, it was less than half full and failed to achieve much more than apply a vague milky sheen to the shapes in the darkness. Most of the time, it was pretty much impossible to see anything, especially on the odd occasions when the trees crowded in. There was no real path to speak of, so I was literally walking blindly ahead, hoping against hope that I wasn't going round in some absurd circle, ending up back where I'd started. Once again my lack of planning returned to haunt me as I cursed myself for forgetting to steal a torch from somewhere before I left.

My stomach rumbled and I looked at my watch. It was almost nine o'clock. OK, I'd got this far without digging into my rations, but it was time to stock up the energy stores. I slumped down against the trunk of a nearby tree and felt in my pack for the container of stew. I pulled it out and took the lid off, letting the heady scent of potatoes waft out. For a moment I almost felt nostalgic for the camp, and this feeling was enhanced by the fact that this was the point at which I realised that I'd completely forgotten to bring any cutlery with me, apart from a small but lethally sharp kitchen knife which was no use whatsoever.

The debacle that followed taught me that, although there are many dishes in the world of cuisine that could reasonably be described as finger food, Belarusian potato stew was definitely not one of them. However, needs must, and ten minutes later I had an empty container and a full stomach. I gave the palms of my hands one last lick, wiped them on the bark of the tree and stood up. Time to move on again.

It was getting cold. I was gambling on my jacket being sufficient to avoid dying of exposure on my first night, on the basis that I'd find somewhere safe to hide out on the following day. However, it would have been even nicer if I'd managed to find somewhere earlier still.

I started walking again. The forest was eerily still. There were no animal sounds at all, not even the tiniest scritch. It was a dead zone. I reckoned that by now I was the only living thing for miles around, and that wasn't necessarily guaranteed for much longer. As I walked, I wondered what was going on back home. Had Dorothy forgiven me yet? If she knew what I was going through, would she be sympathetic? What was she doing now? Was she still trying to find the Vavasor papers, or had she given up for good?

Also, what had happened to Arkady?

I hadn't given much thought to him since being here. My mind went back to that final WhatsApp conversation. I'd been so angry about Matheson faking my messages that I hadn't really taken in what had happened to the others. Arkady wasn't there at the end. He'd gone missing. What was the story behind that? Whose side was he really on? What was his agenda?

I also wondered how Benjamin Unsworth was getting on. At least he'd got to stay in a cushy hotel, although I doubted to some extent whether he would be as convincing playing the role of a renegade mathematician as I was. I really didn't think he would have got past Alexei, so god knows how he was faring with Nikita Petrov. Then again, maybe he'd done enough to allay suspicions while he located Sergei, established what he was up to and buggered off back to Blighty. Job done.

Not for the first time in life, I was beginning to feel that the rest of the world had their destinies sorted out somewhat better than I did. Also, my feet were beginning to get wet. My shoes were intended for strictly urban use only and this part of the forest was distinctly damp underfoot. I began to wonder if it was going to keep dry for the duration of my current ramble. The Belarusian weather had been kind to me so far, but there was no guarantee that it wouldn't decide to turn and dump on me without any warning.

I kept on going. If there was any chance it was going to start raining, I had to find shelter. Somewhere out here there had to be some abandoned settlement I could hang out in. Somewhere that had been evacuated when the reactor blew up. Thinking about this, I got quite excited for a moment, imagining some kind of weird communist-era time capsule, full of gloriously optimistic Soviet posters and long-lost jars of pickled vegetables, and an ancient gas-guzzling monster of a tractor gathering rust in the yard outside.

Then again, the only chance I had of finding anything like that was to literally bump into it. The sky was so dark that I could easily have been sauntering down the high street of Five Year Planville West, population currently zero, without realising it.

As I walked on, my thoughts drifted to how I was going to get safely back to the UK. One possibility was to rock up at the British Embassy (if there was such a thing) in Minsk and explain that I was a passport-less British citizen and could they furnish me with some papers please. There was of course absolutely no possibility that my answers to their follow-up questions ('What is your name?', 'Which flight did you arrive on?', 'What are you actually doing in Belarus?' and so on) would cause any difficulties.

It was about this point when I became aware that my surroundings had subtly changed. A clearing was opening up, and hard lines were beginning to materialise amid the soft, ever-changing shapes of the trees in the darkness. There was a henhouse, a rusty shed with a corrugated-iron roof. And finally, there it was, right in my path. An abandoned wooden cottage. The walls looked as if the slightest passing breeze might blow them over and the paint was peeling off in great chunks, but there was no mistaking it: this had once been someone's home. The only thing missing was the ancient gas-guzzling tractor in the yard.

I almost wept with relief. At last I'd found somewhere I could hunker down for the night, and at this exact moment, the tiredness that I'd been holding at bay all night hit me like a wave and I could barely stand up any more. I went up to the front door and turned the handle. It didn't budge. Perhaps it had seized after all the time it had been abandoned by humans. I tried again, but it was still jammed fast.

I gave up and went to look for a window. There was one just to the left of the front door that looked as if it had swollen slightly and no longer fitted the frame properly. I stuck my fingers under it and tried to pull it outwards. After several minutes heaving, it finally began to move and within a very short space of time, the window was open. I threw my bag in and clambered through, landing on an ancient sofa that had been placed up against it. I reached up and after a couple of attempts managed to jam the window shut again.

The sofa was moth-eaten and threadbare and there were a couple of springs getting ready to poke through into the cushions. But at that precise moment in time, as I lay down on it, it was the softest, downiest bed I had ever been privileged to sleep in. I began to drift off, but then sat up with a start, thinking about bears and other wild animals. I fished in my bag and took out the knife, then I lay down again, clutching it to my chest. I was possibly more tired than I'd ever been in my entire life and within seconds I was away with the fairies.

I awoke with a start in the early hours of the morning. It was only just beginning to get light and when I checked my watch I found it was just after six o'clock. I was about to turn over and go back to sleep when I realised there was a muffled noise coming from somewhere else in the cottage. For half a minute or so I debated with myself as to whether the best policy would be to lie still and hope that whatever it was would go away or if I should get up and investigate.

I decided to lie still and hope that it would go away.

The noise continued.

After another minute or so I decided it was time to rerun the debate, and this time getting up and investigating was the clear winner. I swung my legs off the sofa and, keeping firm hold of the knife, I stood up and walked over to the door. The noise was coming closer and I realised now that it was caused by footsteps walking up and down outside the room. I flattened myself against the wall next to the door and waited for whoever it was to go past.

Surely the Gretzkys couldn't have tracked me down already? I was convinced I'd lost them back at the camp. If they found me here I was surely doomed. They wouldn't even bother to take me back. It would just be a bullet in the head and then my body left for the crows to pick at.

Suddenly the door burst open, hitting me hard on the nose and bouncing back in the face of the intruder.

'Ow!' I yelled, leaping out and brandishing my knife in what I hoped was a threatening manner.

My would-be attacker cried out in alarm, stepping back through the doorway with a hand to his face. I peered at him more closely. Now that I could see him, he didn't look like one of the Gretzky gang. Or if he did, it wasn't one of the bunch I'd seen at the early morning drill. He was quite short, with grey hair and a stubbly beard, and he seemed to be wearing pyjamas.

Still rubbing his nose, the man unleashed a torrent of Belarusian at me, probably along the lines of *What the fuck did you do that for, you clumsy idiot? And by the way, what in the name of Baby Jesus Christ Our Lord and Saviour are you doing in my house?*

No, he definitely wasn't one of the Gretzkys.

'Sorry, what?' I said. 'English.' I wished I'd had more time to prepare a few useful phrases in the local language before I'd come out. I was painfully aware that I was very close to resorting to the time-honoured Brit-abroad trick of speaking very slowly and shouting.

The man muttered 'English' to himself and let loose another stream of Belarusian at me. I imagined this probably boiled down to *That's as may be but being English doesn't excuse you breaking into a stranger's cottage in the middle of the night, you exceptionalist twat.*

'That's right,' I said. 'English.'

'English.' He advanced closer to me. I was still clutching the knife and some residual level of anxiety made me raise it in defence.

'Oh, oh,' he said, putting his hands up. His voice became more conciliatory, and I interpreted his next remarks as

OK, OK, I can see you don't have a clue what to do with that, so you can put that down now. I'm not going to hurt you. Then, unexpectedly, he held out his right hand. 'Artem,' he said, with the beginnings of a smile.

I lowered my knife, transferred it over to my left hand and engaged with the handshake. It was firm but seemed friendly. 'Tom,' I said. I decided to go with my real name, as there seemed absolutely no point in persisting with the Rory thing.

The man held my hand for a while longer, staring into my eyes and then let go. He spoke again, this time more slowly and with a friendly, welcoming manner. He accompanied his speech with gestures around the room, and I imagined him saying *Welcome to my humble abode, wandering stranger. Mi casa es su casa and all that. You may stay here for as long as you like to rest, recuperate and ease your soul.* Of course, it may just as well have been *OK, I will tolerate you here for another five minutes while you gather your shit together, but I catch you still here after that, you're dead meat.*

'Thank you,' I said.

Artem smiled and gave a slight nod. Then he left the room, closing the door behind him. I went back to the couch and sat down. I was still exhausted, and almost without realising it, I lay down and went back to sleep within seconds.

I woke up again at about five minutes past ten, but I did at least feel refreshed. I sat up and looked out of the window. The sun was shining, the henhouse was open and there were chickens strutting around everywhere. Then I remembered the events of the previous night and, with an increasing sinking feeling, I also remembered where I was and, more importantly, how I'd ended up there. Despite having found somewhere to sleep, I was still on the run from the Belarusian mafia.

At that moment, right on cue, I heard the sound of a motor vehicle engine approaching. Some kind of off-road vehicle. A Jeep, perhaps. The vehicle pulled up outside and I heard sounds of car doors slamming and two people shouting at each other. I recognised the voices: Anya and Novikov, and they were heading towards the cottage. I peered out of the window and saw that they were both holding handguns. I ducked back down straight away. Oh shit. This was it. I'd finally reached the end of the road.

Chapter 12

Before they reached the cottage, however, I heard a third voice, coming from the front door, shouting at them in Belarusian. It was Artem and he sounded ballistically angry. His voice had a kind of *Take your fucking Jeep and get the fuck off my fucking property* tone to it. I cringed inwardly, hoping he knew what he was doing. If he'd asked me for my advice, I would have invited him to perhaps consider that this might be a slightly reckless approach to take.

Anya's response was forthright, and included the word 'Gretzky' several times, so I understood her gist to be *Do you know who I am, mate? I'm here on behalf of the Gretzkys. Remember them? Your kids certainly will when they eventually piece together your body.*

Artem seemed completely unfazed by this, and when I risked another glance out of the window, I saw why. He was standing a little way in front of the front door, now fully dressed and maintaining a defiant posture with his arms akimbo. Next to him, an old woman had now materialised. She was dressed almost entirely in black and was even shorter than Artem and she was holding a twin-barrel shotgun in a manner that suggested she was extremely familiar with its use. As far as I could tell from the angle I was watching from, the gun was pointed directly at Anya, who was now engaged in an animated discussion with Novikov.

The exchange of views terminated in Anya holding him back with a gesture that I interpreted as meaning 'Leave these hayseed idiots to me.' She motioned to Novikov to put his gun away and she put hers away too. Then she stepped forward with her arms open and addressed both of them. This time, the name 'Rory Milford' put in several appearances, in close association with the word Английская, which I eventually realised meant 'English'. I guessed her meaning was something along the lines of *OK, sorry for waving the guns around, bit of a misunderstanding there, but if you help us, we promise to leave you alone. We're looking for an English twat called Rory Milford, who's gone and stolen some important papers from us. Although he might not be Rory Milford after all, but to be honest we haven't a clue what his real name is. Can we just come in and have a look around?*

I felt I was beginning to get the hang of this translation lark.

Artem, however, wasn't having any of it. He just said one word: 'не.'

I knew what this meant from my time in the camp. It meant 'no.'

Anya took a step forward, but the old woman advanced to meet her and shoved the shotgun in her face. Anya hesitated, appearing to be weighing up whether her current course of action was worth the risk of having her head blown off and deciding in the circumstances to leave things be for now. She turned back towards the Jeep and said something to Novikov. He nodded and wandered disconsolately off to the other side of the vehicle. He climbed aboard and turned the ignition on. Then she opened the passenger door and paused a moment beside it.

She made one last fiery speech, punctuated by a lot of pointing with her index finger, and my impression was that she was giving them quite the warning: *OK, we'll leave you alone this time, because there's no way even I'm going to shoot an old woman in the face, but any more trouble from you two and you bastards'll be plucked clean for supper.* When she had finished, she whipped out her gun and fired a shot at the chickens, who scattered in all directions, unharmed. The old woman responded by firing both barrels over Anya and Novikov's heads. Then Anya climbed in the Jeep and settled herself in the passenger seat. She closed the door and pointed forwards with her arm. Novikov drove off and we were all left on our own. When I could no longer hear the noise of the engine, I got up and opened the door to my room.

Artem and the old woman were sitting down at a table in the room opposite me, which turned out to be the kitchen. There was a bottle on the table between them, with three small glasses, all of which were filled to the brim.

'Thank you,' I said, hunting around in my brain for the Belarusian word. 'Дзякуй,' I added. It wasn't a commonly used word in the camp, but it was one of the handful I'd managed to pick up.

Artem held up both hands and made what I took to be some deprecatory remarks, including several instances of the word 'Gretzky.' Every time he said this, both he and the old woman spat on the floor. There was clearly some previous between them and the gang. Then he waved his hand towards the old woman and said, 'Бабуля.' Then, with exaggerated emphasis, he added, 'Grandmother.'

She responded with a grimace, then picked up her glass and downed it in one. Artem proceeded to knock back his in one go too, before gesturing to me and the last glass. I looked at my watch. It wasn't even eleven o'clock yet. However, I felt I had to be sociable. So I sat down, picked up my glass and tipped it straight down my neck.

Holy Christ.

My throat was on fire. My stomach was erupting. There were sparks in front of my eyes. I tried to speak, but nothing came out. I had just swallowed a glass of hot lava direct from Mount Etna. My eyes began to water and my cheeks were beginning to melt. I had never drunk anything quite like it, but the worst of it was that I was clearly just about to be obliged to go through it all again.

Artem poured out three more measures and he and the grandmother downed theirs in one easy go. I was still at the point of trying to get my breathing back under control but Artem kept pointing at my glass and saying 'Tom!' over and over again. I didn't appear to have any choice in the matter, so I grasped the glass firmly in my hand and knocked it all back.

Well, nothing mattered now, because I was clearly going to die. There were no two ways about it. My heart had already decided to jump ship and was battering away at my chest trying to make good its escape. I could no longer see anything apart from the vaguest of blurs, my throat had by now been eaten through completely and the liquid fire was flowing freely into any part of my body that took its fancy.

And now I was definitely going to be sick.

I stood up abruptly from the table and staggered over backwards in a heap. I managed to turn myself over and crawled towards the front door, where I managed to drag myself another couple of metres before I could finally hold on no longer and last night's stew spread itself over the ground in front of me. A chicken came over and clucked anxiously at me.

'Fuck off,' I said, and passed out.

I came to on the couch, with Artem standing over me and proffering a cup of something hot. I sat up and gave him a suspicious look.

'Is it safe?'

He nodded and pushed it towards me again. I took it from him and took a sip. It was hot, sweet tea and it brought me back to something close to life within seconds.

'Sorry,' I said.

Artem smiled at me encouragingly and said something which roughly translated as *Get that down you, you hopeless lightweight*. Meanwhile, Grandma was watching me from the doorway, contempt written in every line on her face. And there were a lot of lines on her face. I started to get up, but Artem motioned to me to stay sitting down. Then he went back to the kitchen and returned with a plate of potatoes and eggs and a knife and fork. Then the two of them left me alone.

The eggs were magnificent. I have no idea what the hens found in the forest to feed on, but the yolks were a rich orange colour and tasted of heaven. The potatoes weren't bad either and I wondered if they grew their own vegetables. There certainly didn't seem to be any grocery shops in the immediate vicinity. And my god, I hadn't realised how hungry I was, particularly after losing last night's supper.

When I had finished, I stood up and carried my plate into the kitchen. I was about to offer to clean it myself, when Grandma grabbed it wordlessly from me and stuck it in the sink. I sat down at the table opposite Artem, who was polishing off the rest of his lunch. An ancient radio was plugged into the wall, relaying what sounded like a long and protracted political speech through a speaker that was so tinny you could safely store baked beans in it.

I needed to come up with a plan of action, and that meant making a few calls. With any luck, I reckoned, Artem and his mother must surely have some kind of landline.

'Artem?' I said. 'I need to phone some people.' I made the universal sign by holding my hand to my ear with the thumb and little finger extended.

He replied with a word that I was certain meant something like 'Sure' and produced a state-of-the-art iPhone from his pocket. Well, that was a surprise. He passed it over to me.

'I can pay,' I said. 'I have roubles. A whole wad of roubles.' He looked at me as if he didn't understand, so I got up from the table and went to fetch the money from my bag.

'Ah,' said Artem, when I sat down again and showed him the money. But he waved me away as if he wasn't bothered by such trifles. Eagle-eyed Grandma was less dismissive and wandered over to take a look. She made as if to take some of the cash from me, but Artem put his hand up to stop her.

'They will be international calls,' I said. I pointed with my hand to indicate something far away, but Artem just shrugged. Grandma scowled at me and went back to the washing up.

'Sure?' I said, but Artem was listening to the radio and not paying any attention to me.

I looked at the phone and realised my first problem. I pointed at the screen to draw Artem's attention to it.

'There's no signal,' I said. 'Signal,' I repeated.

'Ah,' said Artem, getting up from the table and beckoning me to follow him. We went outside and, to my alarm, he pointed to the roof of the cottage. He said something and the accompanying gestures made it clear he was telling me that I needed to get up onto his roof. Specifically, I had to get up onto the apex of his roof, assuming of course that it would bear my weight.

On the plus side, it was a single-storey building, but on the downside, if anything went wrong, I could still end up falling quite some distance. I could in fact end up lying significantly injured on the floor of a radioactive forest somewhere in the middle of Belarus, with the added bonus that my only hope of getting help would be for Artem to execute the exact same procedure that had resulted in my injury. I looked around for a ladder, but there didn't seem to be one. I mimed climbing up one to Artem, but he just laughed. It seemed that around these parts, that kind of thing was for wimps.

OK, did I really need to make any of these calls?

Unfortunately, I knew already what the answer to that was.

I put Artem's phone in my pocket, walked over to the window and tried to work out my trajectory to the top. I reached up and grabbed the lintel and swung my right leg up onto the sill. Fortunately, it was only slightly rotten and didn't give completely under my weight. I stopped there for a moment, adjusting my balance. Then I gently pulled my left leg up as well so that I was now positioned in a contorted sideways crouch level with the window itself. I looked back at Artem, who smiled, gave me a thumbs up and disappeared inside.

I felt upwards with my right hand, trying to find my way to the point where the roof joined the wall. I couldn't quite reach, so instead I tried grasping at the gap between one horizontal band of shiplap and the next one up. After a couple of attempts, I managed to find a handhold that was sufficiently strong to enable me to haul myself up far enough to be able to grab the very top of the wall with my left hand.

Bingo. All I had to do now was repeat the procedure with my feet. I lifted my right leg up and searched around for a suitable place at the side of the window to insert it into the woodwork. This took a little longer than it did with my hand but finally I managed to find a place that gave me enough leverage to push away and lift my left leg up while grasping at the roof itself with my right hand. I located a hold above the window for my left foot now and pushed away at that, so that within a moment or two I was balanced with both my hands gripping the roof and the toecaps of my shoes precariously balanced on the shiplap timber of the front wall.

Before gravity had a chance to reassert itself, I pushed on upwards. The roof was tiled, which meant that it had a few more handholds available. I reached up as high as I could possibly go with my right hand and managed to haul my right leg up so that it was now positioned at the junction of the wall and the roof. Very soon, my left leg was up there, too and my body was now sprawled against the slope of the roof. I was now able to shimmy my way to the top and then a little way to the right I located the chimney stack, which was the highest point of the roof.

I had made it.

I made myself as comfortable as I could by straddling the apex of the roof. Then I took out Artem's phone and found to my relief that it now had a very weak signal. I tried to work out where to begin. The problem was, almost all the numbers I needed to call were stored on my own phone, the various constituent parts of which were currently scattered all over the main highway leading away from Minsk International 2 Airport. I racked my brains for an alternative, but however hard I strained, my memory refused to come up with anything else.

Ah, well.

I dialled the number. There was an agonising wait of ten seconds or so and then a voice answered.

'Hello?'

'Hi, Dad. It's me.'

'Bloody hell, son, you sound like you're a thousand miles away.'

'Funnily enough—'

'Look, if it's about replacing the cassette player, I've been looking into this and I have to say you're going to be a bit short of options, particularly if you want to match the reliability of that old Grundig—'

Dorothy and I had managed to destroy my father's old cassette player by attempting to use it to listen to an ancient interview with the Vavasors that we'd come across. It was mostly Dorothy who'd done the destroying bit, out of frustration with its inability to play anything without tying the tape into a massive Gordian knot.

'No, it's not about that—'

'And I hate to be the one to point it out, but if that new young lady of yours had taken a little bit more care with it, we might not be in this position.'

'Dad, I'm not phoning you about the cassette player.'

There was a pause at the other end. 'Well, why the hell not? What am I going to listen to my music on?'

My father had half a dozen cassettes, at least two of which were the best of The Moody Blues and one of those was liable to disappear into the mechanism at the slightest provocation.

'Look,' I said. 'I'm sorry about the cassette player, but we didn't have time to sort it out before we had to go to Greece and now I've ended up in Belarus and I'm using a borrowed phone and I don't have time to talk about that right now.'

On reflection, that sounded a bit brusque. 'Sorry,' I added.

My father guffawed. 'You people and your galivanting around. It's no wonder nothing gets sorted out in the world, is it?'

'What?' I was tempted to ask him at this point what the hell he'd ever done to make the world a better place, but, while it would have been a fair question, it would have been mean of me to ask it, and besides I didn't have the time for a full-blown argument. 'Look, I need a favour, Dad. Remember those alpacas I brought to you to look after a while back?'

'Alpacas?'

'Yes, Dolores and Steven. The ones you palmed off on Mad Dog McFish.'

'Oh, them. What of them?'

'Did you ever phone the woman about them? The woman who actually owns them? I gave you the number. She's called Margot Evercreech.'

'Nah. Haven't had time.'

'What? I... oh, never mind. The important thing is, have you still got the number?'

'The number?'

'Yes. Margot Evercreech's number.'

'Oh, that one. It's on a scrap of paper somewhere.'

'And?'

'Yeah. Been meaning to copy it into my address book.'

'Yes, but can you tell me what it is?'

'Hold on. I'll have to find it first.'

This was followed by roughly five minutes of rummaging sounds, punctuated by yelps from Wally the dog, who was clearly upset at having to get out of the way while my father searched his static caravan for Margot's number.

'Do they do cassette players in – what was it – Belarus?' said my father eventually.

'What?'

149

'Just struck me it's the sort of place they might still make that kind of thing. Might make it a bit easier for you to find one for me. I mean, I suppose I can manage without for a bit longer, but you know.'

'It's not the stone age here, Dad.' I sighed. 'OK, OK, I'll have a look if and when I get back to Minsk. Have you got that number, though?'

'Oh, that. Here you go.' He read it out and I made a note of it on a scrap of paper using one of the pencils I'd taken from the camp.

'Thanks, Dad.'

'You're welcome. Be nice to see you again some time, you know. When you've got a free moment in your busy life.'

'Yes, yes. Look, I've got to go. Talk to you soon.'

I cut the call before I ended up being dragged down another rabbit hole. I felt awful, but at least I was one step closer to my goal. I took a deep breath and called Margot Evercreech's number. She answered almost immediately.

'Evercreech,' she announced.

'Hi, Margot. Look, it's Tom. Tom Winscombe. Last saw you on the mountain in Greece.'

'Ah. That Tom. You stole my alpacas.'

'Yes, about that. My Dad has your number and I've literally just reminded him to call you about returning them.' I was very pleased that I could say this without the slightest hint of a lie.

'I should hope so, too,' she said. 'Well, what do you want?'

'Well, I'm glad you asked me that. It's about Benjamin.'

There was a sharp intake of breath at the other end. 'Benjamin? Is he all right?'

'Oh, sorry, yes he's fine. I just wanted to get hold of him.'

'Hmmm. Might find that a bit of a challenge, Mr Winscombe.'

'Why's that?'

'Well, this is all very curious,' she said. 'Do you remember those awful people who tied us up on the mountain?'

'Helen Matheson and her cronies?'

'That was the name! Yes, do you remember Benjamin saying he used to work for her?'

'Yes. Hard to believe, but—'

'Hard indeed, Mr Winscombe. Well, the most unexpected thing happened on our way home. Can you believe that we bumped into them in Nea Anchialos airport?'

'You're kidding,' I said. Dorothy and I had stayed on for a few days in Meteora after our escape from the mountain. We'd felt we were due a spot of downtime, but I was now slightly disappointed that we'd missed this encounter.

'I assure you I am not,' said Margot Evercreech. 'It was all most awkward, as you can imagine. There we were, booked on the same flight home, less than twenty-four hours after they had us bound hand and foot.'

I restrained myself from pointing out that Margot herself had once held me captive for several hours, mainly because that would have derailed the conversation into further recriminations about the alpacas that I'd ended up taking with me when I'd eventually escaped.

'I have to say,' she continued, 'they didn't look terribly happy about life, especially when I asked them if they'd had a successful trip.'

'Ah well, I can explain that,' I said. 'Remember that document they stole from us? Well, we managed to get it back. Got shredded in a helicopter's blades soon afterwards, though. But still.'

'I can see why that might have made them a little unhappy. Well, as luck would have it, Benjamin ended up sitting next to this Matheson woman on the way home and the long and the short of it is, he ended up back on her books.'

'Weren't you upset, though? He was supposed to be working for you.'

'Depends on how you define work. He was bloody useless as an archivist, that I can tell you. If the truth were told, I was quite pleased to see the back of him. God knows what Matheson sees in him.'

'Cannon fodder, I suspect,' I said.

'Oh dear,' said Margot. 'I hadn't thought of it that way.'

'Well, he's currently in Minsk, pretending to be a mathematician. The same one that I was supposed to be pretending to be, but that's another story. Either way, I need to get in touch with him.'

'I'd better give you his number, then.' She gave it to me and added, 'If you do speak to him, give him my best wishes. He's a nice boy, even if he is completely hopeless. I'd hate to think of him getting into something dangerous.'

I ended the call and looked down at the ground. While I'd been talking, Artem had wandered out again and was staring up at me.

I raised my index finger and shouted down to him, 'One more call, OK?'

He gave me a big thumbs up.

'Or maybe two,' I bellowed.

He didn't seem remotely bothered.

I looked at the screen of his phone and noticed that the battery was getting low. I'd probably be OK for this one, but I might have to get it recharged before the final call. I briefly debated whether I should somehow hand it down to Artem and then wait around up here on the roof until it was charged and he could pass it back up to me or if I should climb all the way down and back up again. Looking up at the sky and the gathering rain clouds, I decided that I'd be better off sheltering inside.

I called Benjamin's number. It rang several times and then went to voicemail: 'Hi, this is Ben... er Doctor Rory... er... Milford. Um. Hi. If you... um... want to leave a message, um... well, yes. Right.'

'Benjamin, you numpty. Answer your phone. This is Tom Winscombe. We need to talk urgently. Don't call my usual number. Call this one.'

I hung up. Trust Benjamin not to answer his bloody phone at a time of crisis. Then I wondered if he was all right. Maybe things weren't going well for him. I felt bad now. I shouldn't have called him a numpty. I was about to give up and climb back down again when the phone suddenly started ringing. I recognised the number as the one I'd just dialled.

'Benjamin?' I said.

'Yeah, hi,' said Benjamin. He was talking in an exaggerated whisper, and the acoustic sounded like he was calling from inside a broom cupboard. 'Can't talk for long. I'm hiding in a broom cupboard. Think they're on to me.'

For once, I suspected that Benjamin was speaking the truth here. His default setting was 'paranoid' at the best of times, but he currently had every right to feel that everyone was after him.

'How come you're still alive, anyway?' he added. 'I thought you were supposed to be dead. That's why I'm impersonating you.'

'Long story,' I said. 'Look, first thing I need from you is Matheson's number.'

'Matheson?'

'Our employer, Benjamin.'

'Oh yeah. Hold on.' I heard sounds of rummaging and then he came back on the line and read out the number. I made a note of it for the next call.

'Next question is: have you found Sergei?'

'No, you see, that's the problem. That's what I've been trying to tell her; it's not that simple.'

'What isn't that simple? Benjamin, you're not making sense.'

'Look, sorry, I've got to go. I need to get out of here before they find me. No time to talk any more.'

Benjamin cut the call. I tried calling him back, but it went straight to voicemail.

Shit. I'd had two objectives for my call with Benjamin: first, to get Matheson's number and second, to find out what was going on with Sergei. Now that I was free of the Gretzkys, my mission was to find him so that I could bring the Vavasor papers back in triumph to Dorothy.

Oh, I suppose I had a third objective: to make sure that Benjamin was all right. In some ways, I felt slightly responsible for whatever happened to him. He wouldn't have got involved if I hadn't allowed myself to get kidnapped.

But out of those three objectives, there was only one that I could say I had unambiguously achieved. I had at least got Matheson's number. I prepared to make the last of my calls, and it was at this point that I realised that the phone in my hand was completely out of battery.

Chapter 13

It took me considerably less time to get down from the roof than it did to get to the top, although my hands were cut to pieces by the time I got there. Artem was busy attending to the chickens.

'Artem,' I said. 'Um...' I handed him his phone. He took it from me, glanced at it and stuffed it away in his jacket pocket.

'Дзякуй,' he said, not realising what I was trying to tell him.

'Well, thank you too,' I said. 'Дзякуй. But what I wanted to say was, I've got a bit of a problem. The phone. It's out of battery. I mean, sorry and all that, but I've got one more call to make and I was just wondering if you'd be able to charge it up for me?'

Artem frowned at me.

'Battery. Flat.' I mimed a vertical hand followed by a horizontal one to indicate phone death.

'Aha!' said Artem, taking out his phone and looking at the display. He pointed to it and repeated my gesture.

'Yes, that's right,' I said. 'Phone dead.'

Artem smiled and put the phone back in his pocket. He clearly wasn't in any great hurry to get me back online. I wondered briefly what would happen if there was a sudden emergency. It would be a little bit tricky trying to charge the phone up at the same time as carrying it up to the roof, after all. I guess they were made of sterner stuff out here in the Exclusion Zone.

I wandered back into the cottage and sat down at the kitchen table. Grandma scowled at me and thrust a bowl of potatoes in my direction. Then she stuck a peeler into my hand. Well, this was at least something I was capable of doing. I'd been practising all week back at the camp.

After a while, Artem came back in and rummaged in a drawer for his iPhone charger. He plugged one end into the phone and one into the wall and for just under a minute, everything went fine. Then the radio suddenly turned off and the phone stopped charging.

Artem sighed and waved a hand in the direction of the forest, while drawing a finger across his neck with the other. I got the impression that the power around here wasn't wholly reliable. I clearly wasn't going to be talking to Helen Matheson for quite some time yet.

I carried on peeling potatoes. Artem busied himself preparing cabbage.

'I was sort of hoping to get to Minsk at some point,' I said after a while. I'd noticed the lack of not only any ancient Soviet tractors but also any vehicular transport whatsoever.

'Minsk?' said Artem, with some incredulity. His subsequent remarks told me that even in this dilapidated cottage in the middle of nowhere, the idea of going to Minsk was quite preposterous.

'I—' Why did I want to go to Minsk anyway? 'I need to find someone who might have something my girlfriend wants,' I said. 'My ex-girlfriend, I should say.'

Artem muttered something very short but sympathetic in Belarusian, which sounded for all the world to me like it meant *Oh, mate.*

In those two imagined words, Artem managed to sum up my entire life as it was at that precise moment. I was in the wrong country, miles away from civilisation, with no communication, no transport and no passport. The only plan I had was to somehow find Sergei and find out what had happened to the Vavasor papers, on the off-chance that this might ingratiate me sufficiently with Dorothy that she would decide to accept me back into her life. But to judge from what Benjamin had managed to say to me during our brief conversation, finding Sergei wasn't going to be at all simple.

Then something struck me that really should have struck me a long time ago. If Sergei had got involved with the Petrovs in the way that Matheson had described, why on earth did they need Rory Milford? Sergei was a world-class mathematician in his own right. After all, he and his brother Maxim had been awarded the Litvinchuk Medallion of Honour for their work. The only explanation was that Sergei had jumped ship already. If he'd ever worked for the Petrovs, he certainly wasn't working for them now.

The other thing that was bothering me was that file at the Institute for Progress and Development. Why did they have a file on Sergei anyway? What was their interest in him, and why did Matheson just happen to have someone working there, too?

Somewhere in all this, there had to be a big picture, but right now the only one I could make out looked as if it was by Jackson Pollock on an off day. But maybe if I got to Minsk and found Benjamin, I might get closer to finding out what was going on.

There was no way around it: I had to get to Minsk.

'Yeah,' I said. 'It's a long, long story, my friend.' But there was no point in revisiting it. There was a limit to the number of imagined conversations I could cope with in one day.

Then Artem pointed to a picture on the wall behind us of a young boy, aged about ten.

'Minsk,' he said.

This seemed an odd thing to say. 'Is that your son?' I said.

'Yes, yes,' said Artem. 'Minsk.'

'Your son is called Minsk?'

'не, не, не!' he said, laughing. 'In Minsk!'

Oh god. Artem was trying to tell me his son lives in Minsk. What an idiot I was. Still, it wasn't any use to me if he lived in Minsk, because I was a couple of hundred miles away from there. But then Artem stood up, went to the wall behind us and took down a calendar. I had almost lost touch with the passing of the days, but when he pointed to one particular day and then to Grandma, myself and him in turn, I realised that it was in fact today. Then he pointed to the same day in the following week and then to the picture of the boy in Minsk.

'Minsk!' he said, pointing out of the window to somewhere far away and then miming someone coming here.

Yes! If I understood him correctly, he was saying that his son was coming home from Minsk next week. Then he stabbed at the next two days in the calendar with his finger and then on the last one, reversed the mime to indicate that the son was going back then.

'Minsk!' he said again.

Bloody hell. He was literally offering me a lift.

'Brilliant!' I said. 'Дзякуй. Дзякуй!'

Artem nodded and said a few words along the lines of *No worries, mate*, while Grandma glowered at the way he was indulging this cuckoo that had blundered into their tidy little nest.

Somehow, I had managed to fall on my feet for once. I was going to get to Minsk after all. I really was. And nothing was going to stop me. Not Anya, not Novikov. Not even the whole of Yuliya Gretzky's private army could stop me now. I was going to Minsk and I was going to find the Petrovs. Then I was going to find Benjamin and after that I was going to find Sergei and then I was finally going to get my hands on the Vavasor papers again. And then everything in my life was going to be wonderful once more.

I spent the next week in the cottage waiting for the son to arrive. Artem and I got along just fine, teaching each other odd bits of each other's language and generally getting stuff done in the way that people who live in a cottage in a remote forest get things done: feeding chickens, growing potatoes, catching and skinning rabbits, repainting the outside walls, that kind of thing. Grandma never said a word to me the whole time, but it may simply have been that she was shy. Either that, or she hated this interloper with every atom, every proton, electron and neutron of her being. Yes, probably the latter, on reflection. Change is always hard to cope with.

Despite the fact that power had been restored and Artem's phone was now fully charged, I didn't get around to calling Helen Matheson. Having spoken to Benjamin and reflecting on Sergei's role in all of this, I decided that talking to her wasn't going to help and in fact it might prove useful for her to think of me as being dead. I did try calling Benjamin again, but it went to voicemail every time and he never called back. I eventually grew tired of climbing up onto the roof in order to check, although I have to say that, having observed Artem on a number of occasions, my mountaineering ability had come on in leaps and bounds.

There were no further visits from the Gretzkys, so I assumed that they'd given up on finding me. From time to time, I took out Alexei's papers from my bag and made a vain attempt to get to grips with them. Whenever I did this, I realised how much I missed Dorothy. She would have made sense of it all. A couple of times I almost got as far as borrowing Artem's phone again just to call her, but every time I went through how the conversation was likely to turn out ('Hi Dorothy, how's tricks? No, don't hang up, I need some help, I mean, basically what it is, I've got some maths here I don't quite understand, I don't suppose you could—'), I abandoned the idea. It wasn't going to do me any good. My only chance of a way back to her was to find Sergei and somehow persuade him to hand over the Vavasor papers.

Grandma's cooking was surprisingly wholesome, given their limited resources, and much more appetising than the food I'd eaten in the Gretzky camp. By the end of the week, I noticed that my belt was becoming tighter. Artem forced me to make another attempt at drinking his home-made vodka and this time, by making a glass last the entire evening, I succeeded in finishing it without any major damage to my internal organs. In the meantime, he and Grandma had each knocked back half a dozen with no apparent ill effect whatsoever, beyond a mild softening of Grandma's scowl whenever she happened to accidentally glance in my direction.

At last the day of junior's arrival dawned and there was a flurry of activity in the cottage. Floors were swept, windows were cleaned, furniture was dusted and polished, and work surfaces were wiped down. If there had been a fatted calf nearby, it would have been given the last rites.

Despite the fact that I was expecting it, the sound of a 4x4 approaching from afar in the late afternoon still caused a moment of panic. But it wasn't Anya and Novikov's Jeep that eventually pulled up outside, but a neat little red Suzuki. The door opened and a smart man of about my own age climbed out. He was wearing a polo shirt, blazer and a pair of artificially distressed jeans, with a pair of completely unnecessary sunglasses balanced on his head. He walked up to the front door carrying a bottle of something colourless and exchanged theatrical hugs with his father and grandmother. Then he turned to me.

'And you must be Mr Tom,' he said in perfect English, with a slight transatlantic accent, 'of whom I have been told so much. I am Mikhail.' He reached out, grabbed my hand and gave it a firm shake.

'Hi Mikhail,' I said.

'It is good to meet you. My father has told me everything.'

'Everything?'

'Everything, my friend,' Mikhail affirmed, patting me on the shoulder. 'Everything.' He stepped back and appraised me in a way that I found disconcerting. 'Funny,' he said. 'I thought you would be bigger, somehow.'

Before I could come up with a witty riposte to this insult, Artem and Grandma turned to go back inside and Mikhail followed them, with me bringing up the tail. Mikhail plonked the bottle, which turned out to be an expensive brand of designer vodka based on silkworms, on the table. Mikhail and Grandma sat down, while Artem went to fetch the glasses.

'Sit,' said Mikhail to me, with an encouraging wave. 'We celebrate!'

'Anything special?' I said, sitting down. I was beginning to feel uneasy about this.

'We celebrate being alive for another day,' said Mikhail. 'Is that not enough?'

I couldn't really argue with this. A glass appeared in front of me, Artem sat down and Mikhail began pouring out the drinks. One by one, the three of them knocked back the stuff in one go. Then they all turned to me. Oh, all right then, I said to myself. What's the worst that can happen?

This was a very different experience to Artem's home brew. It was like one of those bullets that scarcely make a scratch on initial impact but leave an exit wound that stretches halfway across your back. It was indeed silky smooth and went down my throat without giving the sides as much as the barest caress. But when it reached my stomach, some kind of weird chemical reaction went on that let off a firebomb in my guts.

I sat there, stunned, for several minutes waiting for the heat to die down. But it just seemed to keep on building. I opened my mouth to make some inane comment along the lines of 'Well, that was quite something,' but a sudden wave of hot gas lurched upwards and, fearing that there was a decent chance of belching real fire at this point, I clamped my jaws firmly shut.

It was time for round two. I tried putting my hand over my glass, but Mikhail politely moved it away and poured me a glass anyway. Once more, we downed them in one go, but this time it seemed that, since the reaction in my stomach had reached critical mass already, the addition of further quantities of reactant didn't seem to make much difference. Or perhaps I was beginning to get used to it.

We had one more glass each and then Artem and Grandma got up to prepare supper. Mikhail started to get up himself but was gently pushed back by Grandma. The indulgent smile she gave him at this point was a part of her repertoire of facial expressions that hadn't previously made an appearance and I found it more than a little disconcerting.

'So, Tom,' said Mikhail. 'You came here to work for the Gretzkys, I hear.' At the name, Artem and Grandma spat on the floor in perfect unison.

'Well, not exactly,' I said. 'I actually came to work for the Petrovs.'

'Uh-huh.'

'Except not really. I was here undercover. But I got kidnapped at the airport.' Despite Mikhail's brash appearance, I felt I could trust him.

Mikhail frowned. 'Are you some kind of spy, Tom?'

'Not exactly. I work for an agency that does unusual work.'

'Risky work,' said Mikhail.

'Well, I don't know about that. But right now I need to get back to Minsk to find the agent who was sent out to replace me with the Petrovs, so that I can track down the guy we were both sent out here to find.'

'And who is this mysterious guy?'

'His name is Sergei Kravchenko. He's a big deal mathematician.'

Mikhail laughed at this. 'A mathematician, eh? Funnily enough, I have very few contacts within the Belarusian mathematical community.'

'Me neither,' I said.

'But his name is vaguely familiar. Or perhaps I am mistaking him for a reality show star. It is hard to tell sometimes.'

'I don't think he's the sort who'd do that.'

'No. I imagine he is not.' Mikhail sat back and seemed deep in thought for a while.

'Tom,' he said. 'I want you to think very carefully about what I am going to say. At the end of every working day, before I get in my car, I check under it with a mirror. Do you know why I do this, Tom?'

At this point, I realised I had absolutely no idea what Mikhail did for a living. I'd sort of assumed he was a go-ahead business type, with a string of sweatshops somewhere making cheap socks or something.

'What exactly is it you do?' I said.

'Tom, I am a lawyer. I help people who get into disputes with organisations. In my work, I come across people like the Gretzkys and Petrovs of this world all the time. The system here is as corrupt as anywhere, so most of the time I lose. But sometimes I win, and that upsets people. You have been lucky so far, but one day your luck will run out. Look at yourself. Are you really some kind of superhero?'

'I know my limitations,' I said.

'Do you really?'

'So are you saying you're not going to take me back to Minsk with you after all?'

'I'll take you back to Minsk. To the airport. Get on a plane back home and leave us to sort out our own shit.'

'I'll only get a taxi into the city,' I said.

Mikhail gave a sad smile and shook his head. 'Do you have – what is the phrase? – a death wish?' he said.

'Maybe I do. But if I go back home now empty-handed, there'll be nothing for me to go back to.'

At this point, Artem interrupted us with a brief impassioned speech in Belarusian. Mikhail answered him and they had a brief exchange of views, after which he turned back to me with a rueful smile.

'My father says we are being far too serious and solemn. Perhaps he is right.'

I had a bad feeling about what was going to happen next, and right on cue, Mikhail grabbed hold of the vodka bottle. Despite my protestations, he poured us both out another shot. I forced myself to down mine in one go, although this time it didn't seem to have any adverse effects. Perhaps the important bits inside me that might otherwise have put up a complaint had all been neutralised by now.

We continued drinking through the evening meal and well into the night, by which time we had run out of the good stuff and were making serious inroads into Artem's supply of dodgy home brew. I don't remember a lot of what happened, although there was definitely singing involved, and I may also have danced on the table at one point. I liked these people. Even Grandma.

Chapter 14

We all had sore heads the next morning, so we spent the day in relatively light work trying to avoid demanding activities such as conversation. In fact, discourse was largely reduced to grunts and gestures, which at least had the advantage of being inclusive. Even Grandma tried communicating with me at one point, although I may have just misinterpreted a belch.

Just before our evening meal, Mikhail discovered another bottle of the silkworm vodka in the boot of his car, but we didn't even manage to get through half of it.

'So, tomorrow we go back to Minsk,' said Mikhail when we had finished. 'I'll take you to the airport and you'll go home, right?'

'Do I have to?' I said. I didn't want this trip to end in failure. Sure, I'd have one or two interesting stories to tell, but I wouldn't have Dorothy to tell them to.

'It's entirely your choice,' said Mikhail with a sigh. 'Listen, I like you. You have been good company for my father this past week. Think about him. He would be very upset if anything happened to you.'

'It's not going to,' I said.

'Ach, I give up,' said Mikhail, throwing his hands in the air. 'Look, sleep on it and we'll make a final decision in the morning.'

But my mind was made up, and it stayed made up next morning. I packed my things into the back of the 4x4 and said my farewells to Artem and Grandma. It was an unexpectedly emotional moment for all of us. Well, Artem and myself anyway. Grandma waved me away dismissively as I walked to the car, but there was the faintest ghost of a smile as she did so. I felt I had made real progress here.

When Mikhail and I were both seated in the Suzuki, he turned to me and raised an eyebrow.

'Minsk,' I said. 'Please.'

Mikhail just nodded and turned the key in the ignition. The car started and we headed off down the track away from the cottage. After a couple of miles, we came to a wire fence that looked very similar to the ones I'd been through on the way to the Gretzky camp. There was a gate in the middle with a chain and padlock holding it closed. Mikhail passed me a key and I got out and opened the gate for him to drive through. I locked it again and got back in and we drove off once more.

'Why don't they move?' I said.

Mikhail shrugged. 'They like it there. It's all they've known.'

I considered this. Apart from the small problem of the long-term effects of living in a radioactive hot zone and the flaky power and communications, it was in many ways an idyllic spot.

'The government keep promising compensation,' said Mikhail. 'But it never arrives. I do what I can to help. I give my father a phone, but he can't use it without climbing onto the roof, and one day he won't be able to do that. I've told them I would buy them a nice flat in Gomel if they don't want to come all the way to Minsk, but they are – what's the word you use? – pig-headed.'

We'd been driving for about half an hour when the track began to show signs of a rough concrete base and soon after that we hit the main road. We turned right onto it and almost immediately afterwards there was a sign confirming to us that we were quite definitely on the road to Minsk.

'How long until we get there?' I said.

'Just over four hours,' said Mikhail. 'You still sure?'

I didn't say anything.

We travelled on in silence for another half hour, until Mikhail suddenly said, 'I've been thinking.'

I glanced across at him, waiting for him to expand on this.

'The guys who came to the cottage,' he said. 'What were they looking for?'

'Well, me I guess,' I said.

'Yeah, sure, sure. But why would that be?'

'Because I escaped?'

'Sure, but – I mean no offence by this – were you actually any help to them?'

I sighed. 'No, not really,' I said.

There was a long silence.

'Did you steal anything from them, Tom?' said Mikhail eventually.

'What do you mean?'

'Do you have something that belongs to the Gretzkys in your possession?'

'Um… maybe.'

'Thought you might have.'

How did he know?

'Well,' I said. 'It might be some mathematical stuff I stole from them. Load of calculations that I was supposed to be advising the Gretzkys on.'

'Let me guess, it was little Alexei.'

'You know him?'

'I know of him. It's my job to keep up to date with all these guys. Although it's been all change there since the operation in England went wrong.'

I almost nodded in agreement, and then remembered that Mikhail wouldn't have expected me to know about what went down in Pilton Chumpsey. And he certainly wouldn't have expected me to have been there.

'What operation was that?' I said, after a split second.

'Whole load of them got blown up for no good reason as far as we can tell. All very tragic, no doubt, but I can't say I was unhappy to hear about it. So is little Alexei dabbling in complex mathematics now?'

'Apparently so.'

'He's a bright kid, I hear. Well educated.'

'He seemed pretty smart.'

'But, then again, you have all his work.' Mikhail began laughing and shaking his head. 'And that's why you want to find this Sergei guy.'

'Sort of.' I pondered briefly if this was the right time to go into the impact that meeting Sergei might also have on my love life and decided on balance that it wasn't.

'Why did you do this?'

'I'm not sure, really. It sort of happened by accident.'

Mikhail turned and looked at me. 'Do a lot of things happen to you by accident?' he said.

'Now that you come to mention it, yes, I guess they do.'

'I thought so,' said Mikhail. 'You seem the sort of person that accidents happen to. Maybe you should be a bit more careful.'

'Why?' I said.

Mikhail responded by tapping the rear-view mirror. 'Take a look behind us,' he said. 'Don't draw attention to yourself. But you see that Jeep two cars behind?'

I peered surreptitiously out of the back window. I couldn't make anything out.

'Joined us five minutes after we hit the main road,' said Mikhail. 'Been there ever since.'

'Are you sure?' I said.

'I've been trained. You need to spot these things if you want to stay alive. Let's try a little experiment.'

There was a junction coming up soon. Mikhail hung on until the last possible moment and then yanked the steering wheel over. We continued on the minor road for a while, until it became obvious that the two cars immediately behind us had stayed on the main road, while the Jeep had followed us.

'Shit,' I said. 'Could be a coincidence.'

'Could be,' said Mikhail. 'Which is why I'm going to go back onto the main road, just to be sure.'

Very soon we were back on the main road, with the Jeep still on our tail. I could see it clearly now, although it wasn't Anya or anyone else I recognised driving, which was something I guess.

'Coincidence?' said Mikhail.

'Maybe not,' I said. 'So what are we going to do?'

'This,' said Mikhail, pressing his foot firmly on the accelerator. The road we were on still only had one lane in each direction, so the journey suddenly became quite exciting as Mikhail began weaving wildly in and out of the traffic, frequently avoiding oncoming vehicles by the skin of our teeth.

After we'd been doing this for ten minutes or so, Mikhail eased back on the accelerator as we slotted in ahead of a couple of big transporters. There was no sign of our pursuers and I breathed a sigh of relief.

'We're not out of the forest quite yet,' said Mikhail. 'Look.'

There was an extended gap in the traffic coming towards us, and sure enough, before very long the Jeep reappeared, overtaking the two lorries just in time to avoid a white van that was now heading our way.

'They're persistent,' I said.

'Yeah. OK, hang on, I'm going to try something a bit more radical. Make sure you hold onto something.'

I grabbed the door handle with my right hand and the corner of the seat with my left and prepared myself for whatever was about to happen. Meanwhile, Mikhail maintained a steady speed, apparently waiting for an opportunity to take some kind of evasive action. Every time a turning came up on our right, I braced myself ready for him to jerk the steering wheel over. But instead, Mikhail waited until there was a brief gap in the traffic in the opposite lane and then executed a hard left, accelerating into a country lane that I hadn't even spotted, just in time to avoid being crushed by an oncoming oil tanker.

'Jesus,' I said. 'I mean, bloody hell.'

Mikhail didn't respond. He was too busy weaving at high speed through a sequence of single-track roads, while I clung on tight, praying that there wasn't anything coming in the opposite direction. From time to time the SatNav snapped out something that I took to be the Belarusian equivalent of 'Turn around when possible.' Or possibly just 'Heeeeeeeeeeelp!'

This brought back bad memories of the time I'd trashed Lucy's Nissan Micra in the back lanes on the outskirts of Bristol, which ultimately turned out to be the final nail in the coffin of our relationship. Well, that and her preference for Arkady. The difference was on that occasion the threat was entirely imaginary. This time it was all too real.

As we barrelled along, I turned and looked through the rear window from time to time, but there didn't seem to be anyone following us now.

'Can't see anyone,' I said, hoping that Mikhail would take the hint and maybe ease off on the accelerator. But unfortunately, it turned out that he wasn't quite ready to do that yet.

'I mean, I think you've managed to shake them off,' I added. 'Brilliant work. But do you think we need to HOLY FUCK WHAT ARE YOU DOING NOW?'

The reason for my change in tone was the unexpected appearance up ahead of a level crossing, the lights of which had just begun to flash and the barriers of which were just beginning to descend. However, instead of doing what any sane person would have done at this point, which was to jam the brakes on with both feet at once, possibly with the assistance of whoever was sitting in the passenger seat as well, Mikhail chose to accelerate even harder, so that by the time we hit the level crossing – and the word 'hit' is entirely appropriate for what happened to the barriers as we went through – we were going as fast as the engine of the little Suzuki could manage.

As I drew breath on the other side, my ears were assaulted by the horn of the goods train that hurtled through in our wake and had missed us by the slenderest whisker. Mikhail continued onwards but eased off the gas a little now. Surely we were in the clear? All we had to do was find our way back to the main road and we would be safely on our way to Minsk, assuming that the SatNav was still speaking to us. It had gone suspiciously quiet since the level crossing incident.

'My apologies,' said Mikhail. 'That was perhaps a little closer than I would have liked.'

'No, no,' I said. 'It's fine. Fine, fine, fine. I'm good. I mean, we're good. We're all good. Everything's good. Sorry, I'm gibbering. Or am I jabbering? Oh god.' I paused, waiting to see if my heartbeat was planning on returning somewhere close to normal soon. It didn't seem likely. 'Do you think we've lost them now?' I added.

'I hope so.'

'So do I.'

There was a persistent rattling coming from underneath us, and Mikhail muttered something to himself before bringing the car to a stop at the side of the road. He got out and extricated something that might at one time have been part of one of the level crossing barrier arms. He grimaced and hurled it into the field next to us. Then he got back in the car and drove on at a sedate pace for a few more miles. I felt the direction of the car change as we headed back towards the main road. Judging by the position of the sun in the sky, we were describing a large circle, and sure enough after a while a few road signs began to turn up indicating that we would indeed soon be rejoining the main carriageway.

We were a couple of miles away from the junction when the Jeep turned up again. The bastards must have predicted our route and just not bothered even trying to catch up.

'Seen that?' I said.

Mikhail nodded. 'Container in the glove compartment,' he said.

'What?'

'Open up the glove compartment, Tom,' he repeated. There was an edge to it this time, so I quickly did as I was told. The tub was full of tacks and other sharp bits of metal.

Ah, right.

'You know what to do?' said Mikhail.

'Yup,' I said. 'Sure do.'

I lowered my window and leant out. I was just about to scatter the contents of the tub behind us when I noticed that the passenger in the Jeep was also leaning out their window, brandishing some kind of weapon.

'It's OK,' said Mikhail, noting my concern. 'They won't hit us at that range.'

At that exact moment, there was a loud crash followed by a tinkling of glass and a definite breeze coming from the back of the Suzuki.

Mikhail swore in Belarusian and grabbed the tub from me. Steering with his knees, he lowered his own window and with one violent movement, shook every last scrap of metal out of the tub onto the road behind us. I watched through the gaping hole as the Jeep suddenly began to weave backwards and forwards across the road, before coming to a halt in the ditch at the side.

'I didn't want to do that,' said Mikhail as we continued on our way and reached the main carriageway. 'It's preferable to do the minimum amount of damage with these people. There may be repercussions.'

'I'm sorry,' I said.

'It's not your fault.'

'Well, it is. I was the one who stole Alexei's work.'

'Yes, but I was the one who ran away from them. Maybe we should have stopped somewhere public and faced them down. Ah well. At least no one is dead this time.'

'This time?'

'Ha, yes. This time.'

We arrived in Minsk somewhat later than I'd anticipated and considerably colder owing to the open window at the back of the 4x4. We drew up outside the Minsk Metropole and I grabbed my bag from the back.

'I can't thank you enough,' I said, leaning in to say goodbye. 'How can I ever repay you?'

'Maybe one day you will have the opportunity,' said Mikhail.

'Ha,' I said. 'We'll see.'

'Are you sure you want to stay here?' he said, looking up at the hotel. 'It has... connections.'

'I know,' I said. 'That's specifically why I want to stay here.'

'You are a mad man, Tom Winscombe.'

'Maybe I am.'

'Well, if we both survive for long enough, maybe we'll meet up again one day and drink some more and talk all night about how we outwitted the mafia.'

'Maybe we will,' I said.

I stepped away from the Suzuki and waved him off, wondering if I really would ever see him again, and if not, which one of us would be the reason for the failure of our encounter. I headed into the lobby of the Minsk Metropole and marched up to reception.

'I would like a room, please,' I announced. 'For three nights.' I was pretty sure that would be long enough. If I hung around too long, there was a risk that the Gretzkys would track me down and then there'd be all hell to pay. The receptionist gave me a list of prices and, aware of the bulging wad of roubles in my pocket, I abandoned all thoughts of my pension fund and decided to go for one of the more expensive rooms at the top of the hotel, one floor below the penthouse suite. Sod it, I might as well enjoy a bit of luxury while my life was at risk. I checked in under my real name, grabbed my bag and took the lift up to the top. I slumped down onto the bed and breathed a huge sigh of relief. I had made it here at last.

The next thing I had to do was find Benjamin Unsworth. But first, I needed a shower and something decent to eat. I was going to extract every last kopeck of value from Anya's roubles.

*

I slept like a baby that night, properly clean for the first time in weeks and with a stomach full to bursting with the finest food and wine that Minsk could muster. Next morning, I stuffed it even fuller with an excellent buffet breakfast, before heading out for a brief stroll around the block. When I came back, I went up to the reception desk. I needed to find Benjamin Unsworth, or at least his alter ego.

'Can I help you?' said the new receptionist. She was young and blonde, wearing a smart peach-coloured uniform and the name tag 'Svetlana' on her lapel.

'As a matter of fact you can,' I said. 'I'm looking for someone called Dr Rory Milford. I believe he is a resident here.'

A frown flitted across her face before she responded. 'Yes, he is.'

'I wonder if you could tell me which room he is in?'

'I'm afraid I can't do that, sir,' she said.

'Ah. You see, I need to give him something. It's rather important.'

'That is no problem, sir. If you would like to give it to me, I can pass it on to Dr Milford.'

I had anticipated this. 'Ah, that is unfortunate,' I said. 'I need to hand it over to him in person.'

Svetlana put one finger up and said, 'One moment, sir,' before going into deep discussion with one of her colleagues. There was a lot of pointing at me and I got the impression they were trying to work out if I looked like the sort of person it might be safe to let loose on their other guests.

'He is in room 341,' announced Svetlana eventually. Apparently I looked safe after all. This was nice to know.

Full of my new-found vim and energy, I decided to eschew the lift and instead I bounced all the way up to the third floor. I located room 341 and knocked firmly on the door.

'One moment,' came a voice from within. I heard footsteps and then the sounds of the chain being removed. Then the door opened and a dishevelled figure wearing nothing but a flimsy hotel dressing gown blinked out at me. But it wasn't quite the dishevelled figure I'd been expecting.

'Hello?' said Dr Rory Milford.

Chapter 15

I had to improvise fast.

'Ah, Dr Milford?' I said. 'It is Dr Rory Milford, isn't it? I'm from the Embassy. Beam. Jim Beam. I know, I know.' I held out a hand and Milford gave it a limp, reluctant shake. One of these days I would think of a better pseudonym, but it was the best I could come up with on the spur of the moment.

Milford was still staring at me.

'Just wondering if I could have a quick word?' I said. 'Thing is,' I added, leaning in for emphasis and glancing from side to side, 'we've had reports of someone running around Minsk pretending to be you, and the chaps back at HQ wanted to check that everything's all kosher.'

'Um, yes,' said Milford. 'Come in, I guess. Sorry about the state of the place.'

He had reason to apologise. The unmade bed, both chairs and the floor were all festooned with scraps of paper with random equations scribbled over them. The desk itself was reserved for an array of cups and glasses, each with its own residue at the bottom.

'Do you mind?' I said, picking my way towards the windows.

'No, no,' said Milford. 'Go ahead.'

I drew back the curtains and the light of day failed to make much improvement on the scene. I hunted around for somewhere to sit and eventually managed to find a corner of the bed that was clear of mathematical debris. Milford plonked himself down in one of the chairs without bothering to remove the papers lying on it.

'So, then, Dr Milford,' I said.

'Please call me Rory,' said Milford.

'Sure. Do call me Jim, by the way. Everyone else does.' I paused. 'This may sound an odd question, Rory,' I said, 'but did anything unusual happen to you before you came out to Belarus?'

Milford stared at me as if I was some kind of sorcerer.

'Good lord,' he said. 'How did you know?'

I gave a secret smile, as if to indicate that we Foreign Office types had our ways and that I would have to kill him if I told him what they were.

'It was all very strange,' continued Milford. 'I'd just arrived home from my last day at the university before coming here when there was a knock at my door and a couple of what I can only describe as thugs burst in on me.'

'Thugs,' I repeated. I'd surreptitiously stolen one of his sheets of paper and a pencil and I was jotting down odd notes as he talked.

'Yes, and they overpowered me with some sort of chemical – I don't know, do people still use chloroform?'

'I don't know, Rory,' I said. 'Are you sure this really happened?'

'Yes!' said Milford, indignantly. 'And when I woke up I was being held in captivity somewhere.'

'Captivity,' I said, shaking my head as I did so. 'And who do you think might have been responsible for this?'

'It was some woman called Matheson. Claims to run some kind of private agency. You people are probably aware of her.'

I frowned. 'How is that spelt? Does it have a t-h or two t's?'

'How should I know? The only thing I do know is that she shouldn't be going around kidnapping people.'

'And why do you think she did this?' I said.

'I have no idea. But if you say there was someone going around pretending to be me, perhaps he was sent out here by that woman.'

'But why would she do that?'

'I DON'T KNOW.'

'OK, let's move on,' I said. 'How did you escape from this—' here, I checked my notes '—Matheson person?'

'Ah, now that's the interesting bit,' said Milford. 'There was this chap who was supposed to be guarding me, called Brett. Size of a house, he was, but brain the size of a pea. Very easy to manipulate. Took me a couple of weeks, but I eventually won him over. Persuaded him to let me look after the keys for him while he went for a walk outside in the sunshine on the basis that I wouldn't tell the Matheson woman.'

'I see,' I said, trying to sound as if I didn't believe a word of this account, despite the fact that every single word of it rang true. The Brett I remembered really was that thick.

'Anyway, I escaped with just the clothes I was wearing, jumped into a taxi, made my way to Heathrow and hopped on the first available plane to Minsk.'

'I see,' I said again. He was almost certainly telling the truth, but to the untutored eye, it would have all been mightily implausible. 'Where were you being held? Our people in the UK will need to check this out.'

Milford gave me an address in Margate and I made a note of it. I was already looking forward to paying Matheson a visit when all this was over.

'And who is it that you're working for here?' I said.

'Chap called Nikita Petrov. Nice bloke, bit eccentric. Absolutely rolling in cash. You'd be amazed at how much he's prepared to pay me. Funnily enough, he owns this place – did you know that?'

'We're aware of Nikita Petrov, Dr Milford. So what exactly are you doing for him?'

'I'm afraid I've had to sign an NDA,' said Milford.

'What?'

'A non-disclosure agree—'

'I know what it stands for, Dr Milford—'

'Please call me Rory—'

'Dr Milford,' I said, adopting the sternest voice I could muster. 'I have to inform you that Nikita Petrov is a prominent member of the Belarusian mafia. He and his associates are extremely dangerous people and you would be well advised not to have any more dealings with them. I would strongly suggest that you leave the country as soon as possible and decline all further contact with Mr Petrov or any members of his family. I am also hereby impounding all your papers.'

'You can't do th... well, of course, I suppose you can,' said Milford. 'Oh dear.'

I began to gather together everything that was distributed around the room. It had struck me during our conversation that it wasn't hard to dominate proceedings if firstly, the other person somehow imagined you to be a figure of authority and secondly, this other person was dressed only in a flimsy hotel dressing gown, and I fancied pushing it to see how far I could go. Amazingly, Milford didn't even attempt to stop me, even when I asked him to stand up so that I could collect the papers he was sitting on.

I felt a bit bad for bullying this poor man, but it was for his own good. He had no business here. Neither did I, of course, but at least I had a marginally better awareness of what was really going on than Dr Rory Milford.

Once I'd grabbed everything I could find, I walked over to the door.

'I'll be back in half an hour,' I said, turning back towards him. 'I strongly recommend that you will have disappeared by then.'

Milford nodded fervently in agreement.

'By the way,' I said. 'One last thing. Did Petrov ever say anything to you about what happened to the guy who was supposed to be impersonating you?'

'Not a thing,' said Milford. 'Sorry.'

Well, at least he didn't boast about him being brutally murdered, which was something.

I left the room and headed back up to my floor, a large wodge of paper covered in impenetrable mathematics tucked under my arm. I now had the plans of two separate mafia families in my possession, although without anyone to interpret them for me, they might just as well have been extracts from the Voynich manuscript for all the sense they made. If only Dorothy were here with me, life would have been so much easier.

After I'd deposited the Petrovs' papers in my room safe, I jogged down the stairs to the lobby and went out again to clear my head before I made my next move. I was working on the assumption that no one around here knew who I was yet, but it was very likely that this wasn't a sustainable position. I knew from the fact that the Gretzkys had found out about Benjamin's presence in the Metropole that they had spies on the ground here. The fact that I was going under my own name should have muffled the alarm bells for the time being, but the fact that I'd been asking about Milford could easily have set them pinging again.

I wandered around for an hour or so, mulling over different scenarios in my head. Ideally, I still needed to get to Benjamin, if only to make sure he was still alive and also to find out what he knew about Sergei. But I couldn't rely on him. I needed a backup plan, and unfortunately the only one I could think of was to march up to Petrov himself and announce that I was in fact the real (third or fourth) Dr Milford, travelling undercover. Even by my standards, it was a pretty flaky plan.

Back at the hotel, Svetlana called me over as I walked past reception.

'Message for you,' she said. 'From Dr Milford.' She waved an envelope at me. I thanked her, took the letter and opened it as I rode the lift up to the top floor but one. I wasn't sure what to expect, because Milford didn't strike me as the kind of person who'd go to the trouble of sending me a note as he left.

It turned out I was right, because it wasn't from Milford at all. It was from Benjamin Unsworth:

Hope this finds you well and still alive. Tried calling your number several times but went to voicemail every time and I got confused by the voice shouting at me in Belarusian. Had to leave quickly as they've found out I'm not who I say I am. Can't say any more. Don't call me. I'll be standing on the westbound platform in Ploshcha Lyenina metro station every day at 11:15 a.m. if you want to find me. Unsworth.

I looked at my watch. There was just enough time to get there if I left now. I didn't even go to my room. As soon as the lift came to a halt, I pressed the button to descend once more, and within a few minutes I was out in the streets again, heading for Ploshcha Lyenina metro station.

The morning crowds had thinned by the time I reached the station and when I descended onto the platform, there were only a few people hanging around. The platform was a large, airy example of Soviet public architecture: not quite as grand as the Moscow stations I'd seen in pictures, but still impressive. I checked my watch. It was five past eleven.

I tried to look as inconspicuous as possible as I lurked on the station, moving to the westbound side whenever an eastbound train came in and moving over to the eastbound side whenever a westbound train appeared. I'd been doing this for around nine minutes when there was a tap on my shoulder. I spun round to see a woman in a crisp uniform giving me a suspicious look.

The woman said something to me in Belarusian. There were a lot of words in a short space of time and they seemed to form some kind of accusation.

'I'm waiting for my friend,' I said, gesturing towards the westbound platform, where a rush of cold air presaged the arrival of Benjamin's train. This had no effect on the woman, who instead tugged at my sleeve to indicate that I should follow her. I resisted, pointing at the front of the train, which was now entering the far end of the station.

'Friend,' I repeated. 'On this train.'

But the official was having none of it. She gripped my arm and began to drag me away from the platform. I pulled myself away from her, managing to knock her peaked cap off in the process. This got her really angry. She bent down to pick it up, then took out her whistle and gave it three sharp blows. Immediately, a posse of three other railway officials appeared as if from nowhere and surrounded me. The train was now slowing down and I thought I caught sight of Benjamin observing me with some concern from the second carriage from the front.

As the train slowed to a stop, I saw him step out of the carriage just in time to watch me being hustled away past him.

'Look!' I protested, trying to point at him as best I could given the constraints on my arms. 'There he is!' I tried to attract his attention. 'Benjamin!' I bellowed, 'Benjamin! Help!'

But Benjamin had clearly decided that he didn't fancy getting involved in this one. I turned my head and saw him standing on the platform, looking lost and uncertain as to what to do. Then, as the doors were just beginning to close, he dashed back into the train and within a moment he was gone.

Shit.

Well, maybe tomorrow.

Meanwhile, the metro crew were now hauling me up the stairs to the main hall, where they took me into a small office and the woman in charge began some kind of interrogation. I didn't understand a word of it, but I imagined it went something like this:

'Where are you from?'

'England.'

'What are you doing in my country?'

'Tourist.' Anything else would have been way too complicated.

'Why were you lingering in my metro station, like some kind of bad smell?'

'I told you. I was waiting for my friend.'

'Who is this friend?'

'He's called Benjamin. You saw him just now.'

'I didn't see anyone.'

'Well, he made himself scarce. Probably didn't want to be hassled by you. I can sort of see his point.'

'Watch yourself, matey.'

'OK, sorry. Can I go now please?'

There was a long pause. I slipped my hand into my inside pocket and I felt the atmosphere in the room ease a little. I peeled off a few notes and placed them on the table. The interrogator snaffled them up in one slick movement and indicated that I was free to go.

I got up and left the room. Now what the hell was I going to do? I had well and truly cocked up my rendezvous with Benjamin Unsworth and it would be the same time tomorrow before I could try again and even then it might be best to leave it a few days until the heat had died down. I didn't want to run out of roubles, after all.

I decided to spend the rest of the day wandering around Minsk, pretending to do some sightseeing while I decided what I should be doing. Maybe I should just go back to the hotel, grab my things, check out and get the next plane home. But then I'd need my passport to do that and of the two passports that I might have had at my disposal, the fake one was still back in the Gretzkys' camp and the real one was back in the UK. So perhaps I should go and present myself at the British Embassy and try to explain how I'd ended up here with no papers. I could probably invent some theft or accident or something. I was quite sure I could be extremely plausible, although the only proof I had of my identity was an open return plane ticket in the name of Dr Rory Milford and I had just ensured that the real Dr Rory Milford had almost certainly left the country via the airport the very same day.

Well, it was quite a common name, wasn't it?

No, it wasn't, Winscombe. Shut up.

Eventually, at seven o'clock, exhausted and famished, I trudged back to the Metropole, with somewhat less of a spring in my step than I'd had on the way out. As I walked past reception, Svetlana called out to me once again. I feigned surprise and performed a kind of 'What, me? Again?' mime, before I realised that she was looking quite serious and I was looking a bit of a prat.

'Mr Petrov would like to see you now, sir,' she said.

Chapter 16

'Sorry?' I said.

'Mr Petrov. Office on top floor.' She seemed nervous and reluctant to meet my eye.

'Right,' I said. I hadn't realised the man himself had an office in the penthouse on the floor above where I was staying.

I took the lift up to the top floor but one and stopped off at my room to dump my bag and gather my wits together. Had I been rumbled by the Petrovs? Was I about to be accused of being a Gretzky spy? Or maybe this was indeed my opportunity to pretend to be the real Dr Milford. No, I was too tired for that. I was too tired for all of this.

The hell with it.

I stood up, cracked my knuckles together and stepped out of my room. I took the stairs up to the penthouse and found my way barred by a thuggish type wearing a shiny tight-fitting black suit and dark glasses. He didn't appear to have the power of speech, so he just grunted and directed me towards the door. I knocked on it and a voice within barked out something in Belarusian. I opened the door and went in.

The room was a large, spacious office with floor-to-ceiling windows along the length of the external wall. The wall immediately facing me was completely filled with what appeared to be an electronic map of the world. It looked as if at some point in its life, it would have shown daylight and night in real time, but given that Minsk currently appeared to be basking in the midday sunshine, I took it that it had long since broken down. The wall behind me was filled with an array of CCTV screens, displaying what was currently going on in various locations in and around the hotel.

To my right was a desk that was sufficiently big that if you strung a net across its middle, you could have played a decent game of tennis. Behind it sat a man of equally generous proportions, flanked on either side by two more clones of the guard outside. One of the clones was currently engaged in lighting the big man's cigar, which was roughly a foot long. On the wall behind him were a series of photographs of him shaking hands with a series of celebrities.

'Mr Winscombe,' said Nikita Petrov, indicating a chair opposite the desk. 'Do take a seat.' His English was accented, but fluent.

I sat down.

'I hope you are enjoying my little hotel,' he said.

'It's very nice,' I said. Great choice of word when speaking to a mafia boss, Winscombe. Nice.

'Good, good. And you are in which room?'

I gave him my number.

'Ah. Is a good room. We upgrade the taps last year.'

'They're very nice taps,' I said.

'Mine are gold,' said Petrov. 'Investments.'

'Right,' I said. 'Well.'

Petrov paused, as if he was struggling to come up with a precise formulation for his next question. 'Mr Winscombe,' he said eventually. 'I wonder if you can perhaps help me with a mystery?'

'I'd be very happy to,' I said.

Petrov shot me a look that told me to shut the fuck up and leave the fake bonhomie to him, please.

'I am missing something important,' he continued. 'Or rather someone important. A mathematician.'

'Really?'

I got the look again. Shut up, Winscombe.

'His name is Dr Rory Milford and he left my hotel this morning. Shortly after you paid a visit to his room, I believe.'

Ah. So that's why Svetlana wasn't keen on looking me in the eye. She was the one who'd dobbed me in.

'Right,' I said. 'Right. Well.' Right on cue, my powers of improvisation decided it was about time they had some time off. God knows, they'd been working overtime lately. But this still wasn't a good time to be doing my famous imitation of a goldfish. The problem was, my improvising subconscious was all gung-ho for trying out the 'third fake Milford' routine that I'd rehearsed earlier in the day to see if that might buy me a few more minutes, while the conscious part of my brain was frantically trying to clamp down on any such stupidity, knowing the potential cost if it failed.

'Well,' I said again. 'The thing is—'

But I never needed to explain what the thing was, because at that precise moment, there was a loud crash from way down below, sounding as if a consignment of hammers had just made contact with extreme prejudice with the glass in the vicinity of the reception area downstairs. A whole series of alarms now began to ring all over the hotel, including one particularly deafening one in this very room.

'One moment,' said Petrov, who was now looking at the bank of CCTV screens with a frown on his face. I turned to look myself and realised why. The image from the camera in the hotel lobby was currently filled with what I could only describe as a 4x4 motor vehicle. A Jeep, to be precise. All four of us watched in silence as the passenger door swung wide open and a woman in full combat gear wearing sunglasses and carrying an automatic rifle stepped out.

Anya.

She raised her rifle in the air and fired a brief burst up at the ceiling. Plaster dust cascaded down around her in response. Then she slung her rifle over her shoulder and took out a handgun from her pocket. She went over to the reception desk, switching into the scope of a different camera as she did so. She went up to one of Svetlana's colleagues and began making a number of threatening gestures. The man stood his ground, shaking his head. She gestured again and once more he refused. Then she raised her gun, fired one shot, and the man collapsed on the floor.

Anya now moved on to Svetlana. Please, Svetlana, I thought. I know you snitched on me, but please be sensible. Give her what she wants and you might live. It turned out that Svetlana was indeed more sensible than her colleague. She reached into a drawer, took out a whole set of keys and handed them over.

Petrov tutted at her in disgust and then muttered something to one of his sidekicks. The man nodded and, together with his opposite number, they helped Petrov to stand up.

'We will continue this later,' said Petrov, sounding somewhat flustered but still intent on appearing in command of the situation.

'Sure,' I said. 'No problemo.'

Petrov's minders bundled him out of the office, apparently too concerned about their own safety to be bothered about me at this point. I was left sitting in the middle of the room, wondering where I should go next. I decided that it was probably a bad idea for the Gretzkys to find me in the middle of Petrov's penthouse suite, so I got up and left as well. As I reached the stairwell, I saw him and all three of his bodyguards disappearing into the lift.

Avoiding the lifts myself, I scampered down the stairs to see if there was any way I could get out before the place was overrun. When I got to the second floor, I ran along to the fire exit at the end of the corridor to see if it was clear. I gently nudged it open and almost ran into what from the back looked horribly like one of the members of the Gretzky militia.

I stepped carefully back without disturbing him and retraced my steps along the corridor and continued down the stairs towards the ground floor and lobby area. As I turned the last corner, I saw the lifts straight ahead of me and in front of them I saw the prone bodies of Nikita Petrov and his three acolytes. On the ground, checking each of them in turn for signs of life, was Anya.

Trying to avoid breathing or making any sounds louder than the heartbeat of an amoeba, I crept back up the stairs away from the corner and considered my options. I didn't have many right now. There was no way I was going to make my way through the lobby alive, so going down wasn't on the menu. My only choice was to head back up to my room, sit tight and hope that it would all die down before it reached me.

I managed to make my way up to the second floor without incident, but as I passed the entrance to the main corridor, I heard a loud thumping noise coming from there. I flattened myself against the door and looked through the window next to it as a couple of the Gretzky gang began attempting to gain access to one of the rooms. Having got no joy from knocking, they were now using the key. However, the person inside was clearly keen to deny them access, so they were soon forced to resort to kicking at the door instead. The flimsy chunk of medium density fibreboard put up a token resistance before giving way and the two thugs barged their way in. I heard a cry from inside, then a portly middle-aged man wearing nothing apart from a pair of baggy underpants emerged at high speed and disappeared past me down the stairs. This was followed by an exchange of shouts between the two men and sounds of exaggerated rummaging as if their main purpose in entering the room was to find something.

Shit.

Dorothy often used to accuse me of making everything about myself, but this time I really did wonder if I was the cause of all this. Were these people looking for Alexei's papers? If that was the case, I made a mental note that if I ever got out of here alive, I would pay the real Rory Milford another visit. He owed me one. On the other hand, it was getting increasingly likely that I wasn't going to get out alive. I had at least checked in under my own name, but it was only a matter of time before they worked out that the guy with the English-sounding name on the top floor but one might be worth a look. I stepped back just in time as the two members of the Gretzky militia emerged from the room and through the door I heard them smashing in the next room along.

I continued my journey upwards. As I approached the fourth floor, I heard the pounding of boots from a couple of floors above me. I dived into the main corridor to avoid whatever was coming my way and as I peered through the glass into the stairwell I could see another bunch of gunmen on their way down. Once the way was clear again, I rejoined the staircase and carried on my way. Somewhere below me I heard an exchange of gunfire and it seemed that the counter-attack had now begun.

When I got to the tenth floor, I paused for breath and looked out of the window at the back of the building. The rest of Minsk seemed oblivious to what was going on inside the hotel. The traffic still flowed through the streets and barges still sailed down the nearby river. I was trapped in the middle of an internecine war between two rival gangs and right now there didn't seem to be a way out.

I resumed my journey up the stairs and eventually I arrived, exhausted, at the penultimate floor. There hadn't been sight or sound of any weaponry since the fourth floor, so I reckoned I was safe for a little while yet, although what I was going to do with my remaining hours on earth remained uncertain.

I staggered into my room and threw myself on the bed. For want of anything else to do, I turned on the television. The evening news was on, although curiously there was no mention whatsoever of the armed battle currently going on in the Minsk Metropole. Maybe this kind of thing went on all the time here.

After a while I turned the television off, rolled off the bed and went to the safe. I took out the two sets of papers and gathered them together in a single neat pile. Perhaps if I offered them up as a sacrifice, I might myself be spared?

Yeah, right.

Still, it was worth a shot. I flicked through the pages, on the off-chance that some of it might begin to make sense to me. It didn't. So here I was, trapped at the top of a hotel in a country where I couldn't even read the alphabet, while a battle between two mafia factions raged beneath me, with only a few dozen pages of incomprehensible mathematics for company. I really didn't deserve this.

Chapter 17

Just before midnight there was a dull thud that shook the building, followed by a lull in the fighting. Twenty floors below me the guns fell silent and an uneasy, viscous stillness filled the air. I remained crouched awkwardly behind the bed for another couple of minutes to make sure that there was no movement outside before standing up again. I stretched my limbs, cracked my knuckles and took several deep breaths before starting to walk towards the door.

Halfway there, I stopped dead in my tracks, wondering whether I really dared to leave the room and make a bid for freedom down the stairs. There were still no sounds of gunshots from outside, although there were muffled shouts and sounds of engines revving and vehicles moving around. I considered going over to the window to take a look at what was happening but decided against it on the basis that there was too great a chance of being spotted.

Several more minutes went by and I was no nearer to reaching a decision. In fact I went into a kind of frozen trance, which is probably why I didn't register the heavy footsteps thundering down the corridor outside towards my room until it was far too late and I had no time to make any attempt to hide.

There was a crash and a splintering of wood as the door gave way and a man stumbled inside, wild-eyed and waving some kind of automatic weapon above his head. He was shaven-headed with a full moustache and several days' beard growth. I had no idea which faction he belonged to, but right now it hardly mattered. He appeared to struggle for a moment against the force of the momentum that had carried him across the threshold before finally staggering to a halt a few strides in. I stepped a couple of paces back. He swayed from side to side and then seemed to collect himself, bringing his weapon down to my level.

I gaped at him. He peered back at me. Time slowed to a trickle. A kind of rictus crossed his face, then he gave his head a rapid shake, like a dog that has just exited a pond. I tried to speak, but the only thing that succeeded in emerging from my lips was a dry squawk. The man looked as if he was trying to smile, but it came out as more of a wince. Then he pointed the gun in my direction. I raised my hands. He shook his head again, this time very slowly and with great effort. I prepared to die. But he didn't shoot. He just coughed once, spraying tiny flecks of red over his chin. Then he made an odd little noise, halfway between a grunt and a gasp, before dropping his gun and collapsing on the hotel carpet.

I bent down and felt the side of his neck. There was no pulse. Then I noticed the thick flow of blood beginning to ooze out from under him. In all the panic, and not for the first time in my life, I'd missed something rather important.

I stood up again and stepped back from the body, trying to establish whether or not this latest development made my eventual escape from this situation more or less likely. Then again, given that the chances of me escaping before the man's arrival were around one in ten million, it was unlikely that they were going to improve by enough to make it worth betting any substantial amount on.

I pondered briefly as to whether it would be worth my while taking the man's automatic weapon, but it very soon struck me that the amount of damage I could do to myself or indeed any potential rescuer with such a thing far outweighed any advantage it might otherwise have given me. So in the end I just grabbed the two sets of papers and stuffed them under my shirt, Gretzkys' at the front, Petrovs' at the back, then stepped over the body and walked towards the door. Given that there was now no longer any functional lock on it, there was little point in hiding behind it any more.

I peered out into the corridor. There was no one in sight, although there had to be someone else reasonably close at hand, because whatever had happened to my assailant, his wound hadn't been self-inflicted. The carpet had a standard hotel issue red and black pattern on it but as I looked closer, I realised that there were occasional irregularities in it, and as I looked closer these turned out to be spots of blood. I followed them back to where they had come from and eventually ended up at the lift.

On the face of it, this was good news all round. First of all, whoever shot him was almost certainly on a lower floor somewhere. More importantly, the lifts were still working. Moreover, the lift itself hadn't descended yet, so I pressed the button to open the doors. If I could get down to the basement without further incident, I was sure I could find some way of sneaking out through the underground car park. It seemed to me that my chances had just improved to one in five million. On reflection, they would have been better still if I'd borrowed the guy's weapon after all. Even if I'd abandoned it when I left the lift, I could have made use of it in there. But there was no time to go back for it now.

There was a sound like fingernails being dragged down a blackboard and then the doors scraped open. I was about to step in but something stopped me. Something didn't feel right, but it took me a moment to work out exactly what it was. The lift was warmer than it should have been, and there was a slight smell of burning. Then without any warning, there was a series of sharp pings and the lift began to descend, way more rapidly than it should have done.

I stepped sharply back and dropped down to the ground, heart pounding. Still on my hands and knees, I wriggled back towards the lift entrance and peered over into the void, in time to see the lift vanish into a sheet of flame, a dozen or so floors below. Then an ominous metallic whistling sound caused me to haul myself back, only just avoiding the whiplash of the severed cable on its way back up the shaft. I waited another few seconds for it to clear and then risked another glance downwards. There was definitely quite a fire going on down there. My chances of getting out of this alive were heading back down again, well beyond the one in ten million mark.

There was no point hanging around out here. I ran back to my room and picked up the phone. To my relief, there was a dialling tone. I pressed the number to get an outside line. I found the scrap of paper with the number I'd scribbled onto it. I hesitated for an instant, my finger hovering over the keypad. The thing is, I knew it was going to be a complete waste of time, and I'd only end up feeling angry and frustrated. Given that a horrible death was now pretty much inevitable, I should really be spending what time I had left composing myself and reflecting on all the good things that had happened to me.

I tried to reflect.

But I couldn't think of anything good that had happened to me.

Everything was a horrible mess.

And the worst of it was, it was mostly my fault.

I made the call.

The phone rang a couple of times and then a female voice answered.

'Hello? Who is this?'

'It's Tom,' I said. 'Winscombe.'

'Sweetheart,' she said. 'I so didn't expect to hear from you.'

'Please, Matheson,' I said. 'I haven't got much time.'

'Dear me, you do sound agitated. What's happened?'

'I'm trapped at the top of a burning building, that's what's happened.'

'Gosh, how exciting!'

'From where I'm sitting, it's bloody terrifying,' I said. 'But the important thing is, can you please do something to get me out of here?'

There was a long silence. 'Ah,' she said. 'That could be tricky.'

'What do you mean?' I said, trying to remain calm.

'Well, for one thing, that could risk compromising the whole operation. After everything you've been through, too.'

'You have no idea what I've been through.'

'Well, I can imagine you've had a bit of a tough time, sweetheart.'

'It's about to get a lot tougher if I don't get out of here soon.'

'Well, that is true. But the significant thing here, my darling, is that you're currently trapped on the top floor of a burning hotel with no practical means of escape. I can't actually think of any way we could get you out of there, even if we wanted to.'

The worst thing about dealing with her was that she always sounded so reasonable.

'I… isn't there something you can do?' I said. 'There must be something. You've got all the resources of the British government behind you.'

'Ha, if only that were true. Don't forget, sweetheart, we've been privatised. It's all run by the accountants now.'

'But surely you could spare a fucking helicopter or something?'

'Please don't swear, Tom. It's not called for. And where am I going to find a helicopter at this time on a Friday night? You're being silly now.'

'Please!'

'Look, I need to get going. There are some people we need to chase. I really am genuinely sorry about how this has turned out for you. You must be devastated.'

'You have no idea.'

There was a pause. 'No,' she said finally. 'I don't suppose I do.'

'No you f—' I began, before realising that she'd cut the call. I howled with rage and slammed the handset down onto the phone.

Now what?

There was only one thing left to do. It really was the very last resort, but I suppose I had to try.

I took a deep breath and picked the phone up again, relieved that for once I'd managed to override the urge to smash the thing to a thousand pieces. Then I pressed a few buttons and waited while it rang. Eventually another female voice answered and my heart skipped a beat. But it wasn't the voice I was expecting.

'Who is it?' it said. It sounded like Ali.

'Hi,' I said. 'It's me. Tom. Is that—'

I detected a sharp intake of breath at the other end. 'What do you want, fuckwit?' interrupted the voice.

'—Ali?' I said.

'Yeah, Winscombe. The very same. And in case you're wondering, we're working late, because someone fucking has to. I just happened to be passing by her desk while she'd gone to get herself a coffee. I saw an international number come up on her phone, and I thought it might be something important. Won't make that mistake again.'

'Ali, I need help.'

'Fuck off.'

'Seriously, Ali. I'm in a burning hotel in Minsk with no possible means of escape.'

'So why the fuck did you call me?'

'I didn't.'

'Her, then.'

'She's cleverer than I am.'

'So am I, sunshine. So is this earwig that's currently crawling across her desk.' There was a muffled thud. 'Was crawling,' she added, 'and it's still cleverer than you.'

That was unfair, but I felt that it wasn't the right time to pursue the matter. 'Look, do you know anything about escaping from burning buildings?'

'Yeah. LMFGTFU.'

'What?'

'Let me fucking Google that for you.'

'That's not helpful, Ali.'

'It's as helpful as I'm going to be while you're still being fucking lazy about it.'

'Please help me, Ali. Please.'

There was a long pause. Then I heard a muffled altercation going on, with the words 'Spokane' and 'status report' being bandied back and forth. After a while, it sounded as if Ali had now taken her hand off the microphone.

'OK, she's back,' said Ali in a whisper, 'and if she realises that I've nicked her phone and that I'm using it to talk to you, she'll fucking kill me. I'm back at my desk now, so with any luck she won't notice.'

'Put her on,' I said, 'I can explain.'

There was a deep sigh. 'Winscombe,' said Ali, 'of all the fucking terrible ideas you've ever had in your short but painfully underachieving life, that's one of the very worst.'

'Please, Ali.'

'No.'

'It might be the last chance I have of speaking with her.'

'No.'

'I mean, maybe I could try to explain—'

'Jesus, Winscombe,' she hissed, 'can you just shut the fuck up for a minute? Do you want me to save your arse or not?'

I was about to shout back at her, when I realised that something important had changed.

'Sorry?' I said. 'Did you just say—'

'Like I said, Winscombe. Do you want me to get you out of there or not?'

Chapter 18

I considered Ali's question for a nanosecond.

'Of course I want you to get me out of here,' I said.

'Not that you fucking deserve it,' said Ali. 'She's really fucking pissed off.'

'I can imagine.'

'Yeah, well, just shut the fuck up for a minute or two. Got an idea. Just need to… fuck, fuck, fuck not that… fucking hell, no…'

I could scarcely believe that I was actually saying this to myself, but it was good to hear Ali's voice again after all this time. Dorothy's voice would have been even better, and it was quite unbelievably frustrating to know that she was there in the room right next to her. All I needed was to have one quick word and it would all be sorted out.

'OK, look, I've got it confirmed now,' said Ali after a few more minutes of effing and blinding. 'Does the owner of this place you're staying at have some big fuck-off office somewhere near the top of the building?'

'Yes, right above me,' I said. 'I was there earlier on this evening as it happens.'

'What? OK, never mind. Well, according to this list it definitely looks like they sold him one.'

'Slow down, please,' I said. 'Who's they and what have they sold him?'

'Oh, for fuck's sake, Winscombe,' said Ali. 'Keep up. Fucking parachutes is what I'm saying. This company that makes emergency parachutes for wanky rich twats who like an office with a view. Couple of scrotes with an attitude hacked them a while back and chucked their sales list into a pastebin.'

'Sorry, what?' It was taking me a while to process what Ali was telling me.

'What I'm saying, fuckface, is that somewhere in that office above you is a parachute that'll get you out of there. Unless, of course, it's been used already.'

'I don't think so,' I said. 'Last time I saw the boss, he was lying in a pool of blood in the reception area.'

'Well, that's a bonus,' said Ali. 'Right then. That's it. Good luck and next time you're in the area—'

'Yes?'

'—fuck off.'

'Hold on, wait!' I said. 'Does it say where it is? What does it look like?'

'Mate, it's a fucking great parachute backpack thing. Yellow, if that helps.'

The line went dead.

Oh Christ almighty. Was this the only way out?

I really was out of choices now. I went to the door again and checked to make sure there was no one around. Once I was sure the coast was clear, I took the stairs back up to the penthouse.

The first problem was finding the parachute. I scanned the office, but there were no cupboards or anything that would conceal a backpack of any size. There was, however, a door in the opposite corner that led off into a bathroom and next to that there was indeed a decent-sized cupboard with what looked very much like the Belarusian for 'First Aid' written on it. Inevitably, it was firmly locked shut. I needed some kind of lever to jemmy it open with. I raced back into the office, convinced that I was beginning to smell smoke.

I shook each of the drawers in Nikita Petrov's desk in turn, but they were all locked tight. So if there was any sort of jemmy to be found within them, I'd need another jemmy to get at it. There was probably a mathematical name for this sort of problem, and almost certainly Dorothy would have known what it was. Finally, I realised that the chair I'd been sitting on before had a lever to move it up and down. I turned it on its back and set upon it with a wild frenzy until it came free.

Now, armed with my chair lever, I attacked the first aid cabinet and, lo and behold, there was the parachute. There was a moderately alarming notice on it that almost certainly said that it was next due to be maintained three years ago, but I had no time to be quibbling about health and safety regulations now. I put it on and tied the straps to make it secure. Now all I had to do was open the window and fly.

There were, however, no windows I could actually open.

Of course there frigging weren't, because if there were windows you could actually open, someone might actually open one and try to throw themselves out, wouldn't they?

Shit.

What now?

Maybe if I just ran at the glass, it might shatter and I'd go through. There was, of course, a strong possibility that in doing so I might slice off a limb or two in the process but time was running out. I took a deep breath, stepped back as far away from the window as I could and put my head down and charged.

I came to a few moments later, lying on the floor of the office with the headache of my life and the window completely intact. I sat up and tried to orient myself, but my vision was still blurred.

Bollocks. I'd come so far, only to be thwarted at the last. I stood up and wobbled over to where I'd left the chair lever. I picked it up, went over to the window and gave it a good thwack. Absolutely nothing happened. Then I picked up the entire chair and hurled it at the window. Still nothing happened. I kicked out at the window in frustration and nothing continued to happen, apart from the fact that my toe was now throbbing with pain.

OK, Winscombe, come on. Calm down. You always manage to work something out, right?

I took several deep breaths and then I realised how stupid I'd been. My own room downstairs had a window that opened. Sure, it didn't open very far but I was pretty sure that I could use the chair lever to persuade it to open a bit more, maybe even enough for me to climb out. I began to run back towards the door, but when I reached the stairwell, I realised that I'd thought of this several minutes too late, because the flames were now licking round the edges of the floor below. There was no way for me to get back to my room now.

Surely Petrov would have thought of this? If he was sufficiently worried about his safety up here to shell out for his own parachute, he would surely have thought through how he would actually use the thing? I glanced up at the map of the world on the wall opposite me. This was a man who seriously loved his electronic gizmos. Some instinct made me go back to his desk and I started feeling underneath it for buttons. He was just the kind of guy who'd have a whole array of secret buttons, and sure enough there were a dozen of them right there, just where his right hand would have been.

I pressed the first one and the top drawer shot open, hitting me directly amidships and causing me to stumble backwards against the wall behind me, knocking off several pictures as I did so. I picked myself up again and, on observing the contents of the now open drawer, noted that, should I be interested, I now had access to an evil-looking handgun. However, this was low on my priorities right now and I needed to try a few more buttons.

When I pressed the next one, a hole in the ceiling opened, letting a spinning mirror ball drop gently down, to the accompaniment of a seventies disco soundtrack. Generally speaking, I was someone who embraced the idea of working to music, although I was acutely aware on this occasion that 'Stayin' Alive' could turn out to be either prophetic or horribly ironic, depending on how things progressed.

The third button didn't seem to do anything at all, but when I pressed the fourth one, there was a sudden rumbling of machinery from the room next door, followed by a rush of cold night air as the entire sheet of glass in front of me began to slide sideways. Or at least, it slid sideways a couple of feet before there was a horrendous screeching of gears from next door and the whole thing jammed to a halt.

Maintenance clearly wasn't one of Petrov's priorities.

I went up to the gap and assessed my chances of levering myself through it without damaging the parachute on my back. I made one desultory attempt, but it soon became clear that there just wasn't enough space to get myself through. I would have to push the window back another few inches, and the only way I could possibly do that would be to put my back against the side wall and use my legs. This, unfortunately, meant that I would have to take my pack off.

Why was nothing ever simple?

By now, the room was beginning to fill with smoke, which caught the back of my throat, sending me into a coughing fit. I didn't have much time left, so I ripped off the parachute pack and put it down on the floor beside me. Then I leaned back against the wall, trying very, very hard not to look down, because when of course I did accidentally look down, I realised that it was a very, very long way to fall. I closed my eyes, put my foot up against the edge of the window and heaved away. Nothing happened at first and then I suddenly felt it move a fraction of an inch. Encouraged by this, I pushed even harder. Another fraction of an inch. Then a bit more.

And then, without the slightest warning, the window shot away from me and the entire office was now open to the sky. I now found my whole body propelled backwards against the wall, standing one-legged on the threshold, teetering between the floor and oblivion. It was by no means certain which of the two choices my body was going to pick at this point, but in the split second while it was making up its mind, I managed to drop to my knees and grab the parachute pack, just in time for the coin to come down on the side marked oblivion.

As I tumbled out of the window and the cold air rushed past me, I struggled into the parachute harness and fumbled with the fastenings, before giving up and hoping that I could hang on anyway. I found the release ring and gave it a good tug, hoping that at least this was one piece of Petrov kit that actually worked. On present form, I felt there was a decent chance that it would come away in my hand. However, at the very last, my luck turned and I was jerked out of my fall as the parachute unfurled into the night sky above me.

It was, however, by no means certain that I was home and dry. Having avoided certain death from being trapped at the top of a burning building, I was now confronted with the impressively high mortality statistics for BASE jumping, which I remembered reading somewhere was one of the most dangerous recreational activities known to man. I had absolutely no idea of which direction I was heading in, although this didn't really matter as I didn't have the faintest clue as to how I could adjust it either. I experimented with waggling my legs from side to side, much in the manner of an outsize fish that only had limited experience of water, but this didn't seem to have much effect on my trajectory. I just had to hope that I landed somewhere roughly horizontal without too many spikes protruding from it.

The other problem was that, despite the parachute, I was still dropping at an alarming speed, so my ideal landing position would also be one that happened to be lined with mattresses. Unfortunately, it turned out that these were few and far between in Minsk. Instead, all too soon I realised that my descent was about to be brought to an abrupt halt by the top of a two-storey warehouse a couple of streets away from the burning Metropole. I bent my legs into what I assumed was the correct position and prepared for impact.

When I touched down, one ankle twisted awkwardly, before I allowed myself to tip over, continuing to roll sideways, letting the momentum carry me, as per my vague recollections of what paratroopers were supposed to do. This would have worked pretty well if I hadn't forgotten to release the parachute as I landed, and the net result was that I ended up trussed up like some gargantuan alien spider's midnight snack. It took me a good few minutes to extricate myself from the web that I'd managed to construct for myself and by the time I'd located the fire escape and hobbled over to it, there were already a couple of goons with guns climbing up to meet me. It seemed my escape had not gone unnoticed.

At this point I regretted my choice not to bother grabbing the gun from Petrov's desk before taking my leave. I reversed direction as quickly as I could and limped back to the middle of the roof, where I now noticed there was a skylight. I wrenched it open, clung onto the side with both hands and swung around for a while in mid-air, looking for a safe place to drop. There didn't seem to be one, so I let go anyway, hoping that the commando-style roll I'd performed earlier would work here just as well.

In the event, my landing on the upper floor of the warehouse was an exact mirror of my landing on the roof, in that my other ankle was now nicely twisted. At least my limp had become symmetrical. I staggered off towards where I thought the staircase must be, through a series of pallets piled high with bags of flour, until I reached a solid wall blocking my way. The stairs were clearly at the opposite end of the building, and to get there I needed to get past a couple of gun-toting thugs who had just appeared by the lip of the open skylight. I wasn't sure which faction they belonged to, Gretzky or Petrov, and at this point it hardly mattered. I crouched down behind the last pallet in the row to avoid being seen.

I'm not sure if it was just the effects of the elephant-sized dose of adrenalin that was still coursing through my body, or simply the fact that the unprecedented sequence of events that had overtaken me during the past year or so had blunted my reactions to life-threatening situations. Either way, I was still surprised that my response to my present predicament didn't involve collapsing into a blubbering heap. Instead, it was more a case of, well, I'd got this far and I wasn't going to let a couple of numbskulls stop me now. No way.

While the two of them were engaged in getting safely down from the ceiling, I noticed that next to where I was hiding there was a kind of control unit, with buttons that appeared to indicate forwards and backwards, left and right, and up and down. I gave the forwards one a gentle nudge and a crane in the ceiling moved accordingly. I used the controls to move the crane so that it was positioned directly above one of the pallets and lowered it, fairground-game style, onto the top bag of flour. As the two mafiosi landed on the ground and picked themselves up again, I found another button on the control panel that activated the grabber on the end of the crane and I managed, after a number of attempts, to get hold of a large bag of flour. I picked it up and steered it to a position over where they were standing and let it drop before they realised what was going on.

Both of them went down in a flurry of white dust and coughing, flapping their arms and making things worse with every move. Meanwhile, I located the nearest pallet to them and repeated the manoeuvre, then once more for luck, until I judged it safe to make a run, or at least as quick a stagger as I could manage, for it past them. As I reached the staircase at the far end of the warehouse, a single shot ricocheted off the wall in front of me, but it was a wild shot and missed me by several feet. I hopped down the stairs, found the front door of the building and lurched out into the street.

The late night crowd had largely dissipated by now, and those who remained all seemed to be heading towards the Metropole, drawn like moths to the literal flames of a burning building. The air was full of the noise of fire engines heading towards the hotel from all corners of the city. I decided to head in the opposite direction. I'd seen enough destruction for one night and I was tired. I reckoned that by the time my attackers had dusted themselves off sufficiently to be able to make an appearance, I would be long gone from the area.

Without any real sense of where I was going, I limped on into the night and after a while I found myself by the side of a river. I trudged on into the night until I came across a boatyard, where I could see several small pleasure craft that had been hauled up out of the water and put onto stands for maintenance work. As soon as I saw these, I knew what I had to do.

The gate to the yard was protected by a padlock, but the chain link fence on either side wasn't particularly high and there wasn't anyone around to observe me. Even with my ankles in their present state, it was a relatively simple job to get myself inside. Once I was there, all that remained for me to do was find somewhere to spend what was left of the night. I chose a small cruiser called 'мілая аня', partly because it looked spacious enough for me to bed down in but mainly because it was the first one I came to.

There was a ladder lying on the ground nearby, and when I placed it against the side of the boat, it reached to a couple of feet under the guard rail. I gave the boat a good push to check that it was secure on its mounting blocks and then climbed up. Once I was at the top, I reached up and swung one aching leg onto the top of the rail and then pushed myself over, landing in an awkward heap on the deck. From there I negotiated my passage down the gangway into the forward berth. The bed had been stripped, but the mattress was still inviting. I lay down on it and was fast asleep within five seconds.

Chapter 19

I was awoken by sounds of hammers banging and shouts from people moving around. I looked at my watch and discovered that it was already half past nine. I risked a glance out of the window and saw that it was a crisp autumn morning, with a fine mist wafting in from the river. I sniffed the air, smelled burning and panicked for a moment until I realised the smell was coming from myself. I swung my legs over the side of the bed and realised that my ankles had locked solid during the night, having swollen to the size of cantaloupe melons. I tried rubbing them but this only seemed to make things worse.

I listened to the direction of the sounds outside and as far as I could tell, there weren't any coming from my immediate vicinity, so it seemed that the owner of this particular vessel was having a day off. In fact, judging from the state of the boat, whoever owned it hadn't actually got round to doing any kind of maintenance work for several years. If I played my cards right, I could hunker down here for several days until the heat had died down.

I would, however, need to get out at some point soon to get something to eat. I was, I realised, quite unbelievably hungry. I removed the two now extremely crinkled sets of mathematical papers from beneath my shirt and secreted them underneath the mattress. Then I stood up and staggered awkwardly around the bed to get to the gangway. Given the state of my ankles, my preferred method of ambulation involved rocking awkwardly from side to side, and anything involving going up and down stairs was fraught with difficulty. When I reached the deck, I dropped to my knees and crawled over to the side, taking care not to be seen, and at this point I realised my first problem. Someone had borrowed my ladder.

I looked down and estimated the distance to the ground. It was comparable to the distance I had jumped from the warehouse skylight, but I'd still had one good ankle then. If I landed on either of these crocks, the damage could be permanent. It seemed that I had succeeded in stranding myself.

I could of course have shouted for help, and I have no doubt that I would have been rescued. However, that would have blown my cover and led to all sorts of awkward questions being asked and I was still too mentally tired to have to deal with all that stuff. Then I remembered that the one thing that boats always had loads of lying around was rope. Surely there would be some kind of anchor at the very least? I listened again to the sounds of hammering and judged that the boatyard was clear on the port side of the boat. So I continued crawling on my hands and knees past the wheelhouse and towards the prow, where the anchor mechanism was located inside a hatch.

The first problem was that the anchor was controlled by an electronic system, and it very soon became apparent that there was nothing left in the boat's battery. There was, however, a big red lever that looked very much as if it was something you might pull in the event of an emergency if you happened to find yourself with a flat battery in a situation where you urgently needed to stop drifting. It had some impressively dire warnings on it about being only for use in an emergency, which made it even more promising. I grabbed hold of it and gave it a good heave, at which point I heard a metallic slithering sound as several metres of steel hawser spun out, punctuated by a clang as it reached the ground which was of a similar volume to one that might be put to good use if the imminent arrival of an apocalypse needed to be announced. The sound echoed on for way too long around the boatyard to let everyone who might have been interested know that there was a stowaway on board.

The hammering stopped, and there was a brief exchange of shouts in Belarusian. I imagined it went something like this.

'Hey, did you hear that? Either Vadim's boat's been possessed by a demon or someone's snuck on board.'

'Probably some kid finding somewhere to get jiggy with his girlfriend.'

'Do you think we ought to do something?'

'Nah. Vadim's a twat. Fuck him.'

Because nothing more happened and the banging started up again soon afterwards. It seemed no one was that bothered about investigating any potential interlopers, although I imagined that this might change if they happened to spot such an interloper attempting to shin down the anchor cable. I would have to pick my moment quite carefully, and take whatever steps were necessary to stifle the inevitable screams of pain arising from the use of my cantaloupe melon ankles as brakes.

The hell with it. I needed food. I scanned the area where the voices had come from and established that they were indeed the only two people currently working in the yard. Once I was sure that they both had their backs to me and that they were thoroughly engrossed in their work, I lowered myself over the prow and clung onto the anchor cable with my legs as best I could, while keeping as tight a grip with my fists as I possibly could.

The cable clearly wasn't intended for human use and it wasn't so much my ankles that were the problem now but my hands. By the time I reached the bottom, they were sufficiently full of metallic swarf that you would have got a decent price for my upper limbs at a scrap metal merchant's. I tried brushing them together to clean myself up but only succeeded in ridding myself of half of the shards, embedding the rest even further under my skin.

At least the gate to the yard was open now, so I didn't have to climb over the fence. I found a mini supermarket close by and bought a selection of smoked meats along with several packs of rolls that looked spectacularly unappetising but also appeared to be sufficiently packed with preservatives that they would last me for the next few months if necessary. I also threw in a few bananas and several bottles of water. Finally, after giving one of the staff a flamboyant performance of my mime entitled 'I am in excruciating pain', she nervously directed me to a shelf full of medicines, where I located some tablets that looked as if they would get a horse moving again, along with some cream that was probably for nappy rash but would also go some way towards soothing my stinging hands.

I found myself a bench to sit on close to the river and washed two of the tablets down with most of one of the bottles of water. Then I made myself an impromptu sandwich by tearing one of the rolls apart and stuffing it with a slice of sausage. I took a large bite out of it and almost instantly started to feel sane again. I also realised that since my escape the previous night my brain had been in such a state that I hadn't actually thought about what I was going to do. But as soon as the painkillers and nutrients began to work their way into my bloodstream, my IQ seemed to rise by a good fifty per cent and I began to feel capable of putting together some sort of plan again.

My first objective was – obviously – to stay alive. In order to do this, I had to avoid any Petrovs who were still around, along with whichever members of the Gretzky raiding party were still at large in the capital until such time as they gave up and went home, at least while I was incapable of running away from any potential encounter with them. I guess it was possible that representatives from either group would be checking locations such as the boatyard for hiding places, and perhaps they even had their own local networks of informers who might tip them off about unusual events, such as demonically possessed anchors suddenly falling out of dry-docked cruisers. Setting all those possibilities to one side, on the whole I felt I was as safe as I could possibly be.

Once I felt ready to step into the outside world, my initial port of call was once again, I realised with a sinking heart, Benjamin Unsworth. Whatever misgivings I may have had about him, he was still the key to finding Sergei. I would have to lurk around Ploshcha Lyenina metro station at the designated time and hope that I managed to avoid looking suspicious this time.

I hadn't really considered what I was going to do if and when I tracked Sergei down. In fact, I hadn't really thought much about Sergei anyway, other than as a conduit to the elusive Vavasor papers. What had he been doing with the Petrovs? Why had he gone rogue? Whose side was he really on?

Finally, how on earth was I ever going to get home again? My life back there was well and truly wrecked now, but what was left was still a life of a sort. I finished my sandwich and sat for a while watching the river flow gently past. An unexpected thought drifted into my head: perhaps I could abandon it all and build a new life here, with someone like Svetlana from the Metropole. God, I hoped she'd got out of there alive. I felt instantly guilty for not having thought about her up until now.

What a godawful mess it all was.

I stood up, picked up the carrier bags from the minimart and hobbled back to the boatyard. I'd hoped that a walk might have gone some way towards loosening things up, but my ankles were feeling no better for their outing. In fact, in many ways they were even stiffer than before and I was having to walk in a series of weird arcs, like a crab trying to infiltrate a convention of lobsters. The gate to the yard was still open, so I went in, trying to look as nonchalant as possible, given my unnatural gait and the large number of bags of shopping that I was laden down with. I was now faced with the reverse of this morning's problem; how to get back on board ship without drawing attention to myself.

There was obviously no way I was going to even contemplate shinning up the anchor cable. Equally, I wasn't about to chance borrowing a ladder. In the daylight, however, I now noticed that there was another abandoned cabin cruiser of around the same size as the one I'd spent the previous night in, lying at a forty-five degree angle on its side on the far side of the yard. As a temporary home, I would have preferred something with a slightly better sense of the horizontal, but beggars couldn't be choosers. I made my way over to it, careful once again to avoid attracting the attention of the guys with the hammers, and found my way in.

There was still water sloshing around in the bottom of this one, although on reflection, it was probably only rainwater that had leaked in. I managed to find myself a place to sit in the corner of the wheelhouse and after a few experiments to determine whether it was preferable to have my legs on the side and my back on the deck or vice versa, I decided to plant one leg on the side and one on the deck, with my back up against the front of the wheelhouse.

Over the next week, I made myself a home there. I spent the first evening bailing out and fixing up a tarpaulin that I'd stolen from one of the other boats as a covering against the weather. On the second night, I went back to the boat I'd spent the first night in and salvaged the mattress, along with the mathematical papers, which were now nicely ironed flat again. I laid the mattress on the interior side of the hull, where the angle that the boat was lying at meant it was close enough to horizontal. On the third night, I found some sail material lying around that I managed to turn into some decent bed sheets. In fact, by the end of the week, I was all ready to pick out curtain material.

Meanwhile, the swelling in my ankles was beginning to subside and I was beginning to feel somewhere close to human again. Soon I would be ready to face the world and resume my search for Sergei and the Vavasor papers and once that was over I could finally go home and sort out my life.

On the morning of the eighth day in the boatyard, I woke up fresh, stretched my legs and made myself my usual breakfast of preservative-enhanced roll filled with salami. Then I took a few deep breaths and waggled my legs around. I looked at my watch. It was just after ten o'clock. I had just enough time to get to Ploshcha Lyenina in order to meet my scheduled connection with Benjamin Unsworth.

I took the Gretzky and Petrov papers out from underneath the mattress and stuffed them back under my shirt. Then I gathered together the week's rubbish and dumped it in a skip that I passed along the way. Whichever direction today's events took, I didn't anticipate returning to my temporary home again. It was a fine, crisp, sunny autumn day as I made my way to the metro station and it was good to be out and about again, even if I had a constant feeling that I was being followed. I reassured myself that I was just being paranoid. I'd spent too much time on my own lately with nothing but my own thoughts for company.

Halfway there and I quite definitely was being followed. There were a couple of them, maybe fifty metres behind me, trying too hard not to be noticed. I didn't recognise them, so it was hard to tell which faction they belonged to. Not that it really made a lot of difference. I sped up. They sped up. I slowed down. They slowed down too. I wondered when they were likely to make their move, and if so, what form that move might take. Where they intending to bundle me into a car, perhaps? Or just shoot me with some kind of evil poisoned dart?

Whatever it was that they had planned, they didn't do anything before we reached the metro station. I vaguely wondered about trying something clever like going down one side of a set of stairs, leaping over the barrier and doubling back up the other side, but if I did that, I wouldn't be able to make my connection with Benjamin, and that, after all, was the whole point of this.

I stuck to my original plan and very soon I found myself on the westbound platform again. It was ten past eleven. I just had to stay out of trouble for five minutes, whether that trouble happened to come from the people following me or from the metro officials, who I assumed were even now wondering if that guy lurking down there looked at all familiar.

A train pulled in on the westbound side. A dozen or so people got off and the rest of the people on the platform apart from myself got on. The train pulled out again and I suddenly found myself on an empty platform. I looked over to the stairs and saw my two pursuers at the bottom, walking purposefully in my direction. I started to move away from them and they immediately began to run. But my legs still weren't back to normal yet and I realised that they were much faster than me.

I kept on going, but then I made the fatal mistake of looking back to see where they were, and at that moment I ran into a row of benches that some idiot had decided to plant in the middle of the platform. My right leg caught on the edge of the frontmost seat while my left leg strode enthusiastically onwards. After a brief struggle, I managed to extricate my right leg, but my left was still propelling my body forwards, reduced to hopping now. For an instant, my centre of gravity seemed undecided as to where it felt most comfortable, although after some thought, it came down on the side of somewhere a little to the left and also a bit lower. Much, much lower, in fact.

I put my left hand out to stop my fall and flailed about with my right, trying to grab on to the illuminated sign in the middle of the benches and missing completely. Eventually, my body hit the ground and all forward motion stopped for an instant. I gripped the last bench with my right hand and tried to pull myself up at the same time as my legs were attempting to get running again. Unfortunately, they didn't wait until the rest of my body was vertical, with the result that I lurched on for half a dozen paces before losing control once again.

By this time, however, my pursuers had caught up with me. I turned over on my back to see the pair of them looking down at me, both with guns in their hands pointing directly at me. Where were the metro staff? How much had they been paid to look the other way?

'Where are they?' said one of them.

'Where are what?' I said.

'You know,' said the other one, kneeling down and pressing the gun into my face.

'I genuinely don't,' I said. 'Who are you?'

But he didn't have a chance to tell me, because at that precise moment, his colleague collapsed on top of him, knocking the gun out of his hand. A large hand stretched out, grabbed hold of my arm and jerked me to my feet, dragging me towards the eastbound train that was just arriving.

I looked up at the mountain of a man who had just saved my life.

'Come with me,' he said. 'Need to get out of here.'

'Hold on,' I said. 'I have to get the westbound train. I'm meeting Benjamin Unsworth. He can tell me where Sergei is.'

'Sergei dead,' said Arkady. 'Sorry.'

Chapter 20

'Arkady?' I said. 'Is it really you?'

'Yes, is me.'

'Sergei is dead?'

'Yes,' said Arkady. 'Is sad. Especially for Carla.'

'Who's Carla?' I vaguely remembered the name.

'Is girlfriend.'

'Ah, yes.' The metro train rumbled on into the darkness. 'Where are we going now?' I said.

'I tell you later. We don't speak now.'

'Is Benjamin all right?'

'Benjamin safe. On way to England. No more talk, please.'

Eventually the train reached Уручча and Arkady announced, 'End of line. We get off now.'

Outside the metro station, there was a line of cars parked, facing inwards.

'Follow me,' said Arkady, indicating that we should turn left. We walked on together for a few hundred metres until we came across a car that I recognised: Arkady's ancient Škoda.

'Hey,' I said. 'You've still got it.' The last time I'd seen that car, I was driving it away after saving Dorothy while unwittingly delivering the bomb that had blown up the UK branch of the Gretzky family.

'Is good, no?' he said, patting it lovingly on the bonnet.

'Is very good, Arkady,' I said.

Arkady reversed out into the road and we drove off in silence.

'Where are we going?' I said again. 'Actually,' I added, 'scrub that. First of all, tell me what you're doing here.'

'Is long story.'

'Is it a long journey?' I said.

Arkady gave a sad smile. 'OK, is long enough for story.' He paused for a moment, seeming to gather his thoughts. 'First of all,' he said, 'I have to apologise.'

'Why?' Right now, I felt Arkady had very little indeed to apologise for. If it hadn't been for his intervention just now, I would have almost certainly ended up dumped on the metro track with several more orifices in my body than I'd started the day with and significantly less blood.

'I tell you lie,' he said. 'About Institute for Progress and Development.'

'Oh,' I said.

'I make the cock and bullshit story.'

'Oh.'

'Is my fault, sorry.'

'Right. So, hold on, Arkady. What exactly did you lie about?'

'Sergei ask me to break in to Institute.'

'Sergei? I thought this was all about the Vavasor papers.'

'No. Is about Sergei.'

'Oh.'

Arkady took a deep breath and was about to say something, but he had to swerve to avoid an oncoming articulated lorry. He swore at it in Belarusian and blew his horn several times.

'Sorry,' he said. 'So. Sergei.' Another deep breath. 'Sergei is difficult man. I think your word is impetuous. He very clever but he also wild. Understand he can also be angry man. So when Gretzky people kill his little brother he get mad.'

'I can completely understand why,' I said.

'This true. So when I call him to help free your Dorothy from Gretzky people, he see opportunity. He make bomb and you deliver bomb. Understand he did not mean to kill you.'

'But he might have done.'

'Is true. This why he make sure you not trigger bomb until you hand over to Gretzky.'

'He changed the combination on the lock, yes.' I remembered all of this. It was one of the most tense moments of my entire life.

'So bomb blow up, you and your Dorothy escape and you live happy after.'

'Well up to a point—'

'But Sergei in danger now. I also in danger because I know Sergei.'

'Were there other Gretzky members in England at the time?'

'Is good question. There are people with sympathies, OK?'

Arkady turned to me and gave me a meaningful look. He was interrupted by a loud horn from the opposite lane and he jerked back just in time to avoid a truck carrying a load of sheep.

'But who would be sympathetic to an organisation like the Gretzkys?' I said.

'I think you know, Tom.'

'Are you talking about people like Rufus Fairbanks?'

'Who is Fairbanks?' said Arkady. 'I recognise name.'

'He is – was – an investment banker who helped them launder their profits. With the help of the Vavasor twins.'

'Ah, the Vavasors. Always the Vavasors, eh?'

'I know what you mean,' I said. Everything always seemed to come back to them.

'I do not know much of this Fairbanks,' said Arkady. 'But there are many in the London City who are close to Gretzkys. Also people connected to Institute for Progress and Development.'

'The Institute? You're kidding. But that would put them on the same side as Fairbanks, and they were trying to rip him off with their cryptocurrency scam.'

'Tom, listen to me. There are no sides. Only chaos. This most important thing you hear today. Please remember this.'

'Chaos? As in real-life chaos or mathematical chaos?'

'Is same thing.'

There was a long silence between us. I waited for Arkady to say something.

'So, about Sergei?' I said eventually.

'Yes, Sergei. Sorry. Sergei has blown up the Gretzkys and he also now has Vavasor papers.'

'So they definitely didn't get blown up? You were telling the truth about that at least.'

'No. Sergei good mathematician. He not want to destroy work of others. Even if they help Gretzkys.'

'Fair enough. So what did he do next? What did you do next, Arkady?'

'Sergei leave England and go back to Belarus. I stay in England, but I find place to hide. Is sad because Lucy not like living like this and she leave me.'

I'd temporarily forgotten that he was the one who'd elbowed me out of the way to get in with her. I guess I really would have to forgive him now that he had saved my life, although I had to admit that it still rankled.

'Why did Sergei come back here, though? Surely it can't have been safe for him.'

'Safer than England. Here he has friends. In England is small community. Everyone know everything about everyone else. As long as Sergei avoid Gretzkys, Sergei safe.'

'So what happened then?'

'For a while, everything fine. Then one day, Sergei come home to find flat in terrible mess and Vavasor papers gone.'

'I thought you said that happened in London. Papers hidden under a mattress and stuff.'

'Remember, Tom. I said it was cock and bullshit story. Sorry.'

'Oh. So did he go to the police?'

'Ha, no. That would be dangerous. No, Sergei has list of suspects. First, Isaac Vavasor. This fair enough. Papers are his and Sergei intend to return them soon anyway. But Isaac busy with other things, and not interested in papers for now. Second, the Gretzkys.'

'I don't think they have the papers,' I said. 'I'm sure Alexei would have said if he had.'

'Is true. Sergei think so too. He think if Gretzkys know where he live, they more interested in killing him than papers.'

'Well, there's a certain logic to that.'

'Third, the Petrovs. Is definite possibility. They hear rumour about what Gretzkys do with chaos. They want to do same.'

'I can see that.'

'Fourth, this man Julian Gowers.'

'Who?' The name Gowers rang a bell, but I couldn't place him quite yet.

'He work for Institute for Progress and Development. Is clever man. He very keen on chaos.'

That was it. He was the guy mentioned on the Post-It note attached to Sergei's file. FAO Gowers. For the attention of Julian Gowers.

'But why?' I said. 'For whose benefit? What's so good about chaos? This is what I don't understand. These Institute people, aren't they backed by big business? Isn't business all about stability?' Having spent some time acting as a public relations manager for a few businesses, I reckoned I knew a bit about these things.

'There is English word for you, Tom, but I cannot pronounce. Begins with N, sounds like "knife".'

'Naïve?' It wouldn't have been the first time I'd had that adjective applied to me.

'Ha, yes,' said Arkady. 'No, Tom, is not about stability. Is about chaos. And is not business, is billionaires. Business just a toy. People just toys, too. Billionaires create chaos, billionaires make money.'

'Disaster capitalism?' I said. I'd heard the phrase bandied around, and for the first time I was beginning to see what it meant.

'That is what is called.'

'So what did Sergei do?'

'Sergei go to Nikita Petrov.'

'Petrov?!' I exclaimed. 'But wouldn't that have been even more dangerous than going to the police?'

'I tell you Sergei impetuous. But was a clever idea. He say, "Look Mr Petrov, I am great mathematician. I can bring down Gretzkys for you."'

'But how could he do that?'

'Chaos, Tom. Chaos. Remember the word.'

'OK, OK, but how?'

'First, get data. Where does money come from? Where does it go? How much of this? How much of that? Second, find pattern. Inter... inter... what is word? Interactions. Third, find lever for chaos. Fourth, pull lever.'

'Bloody hell. So are you saying that Sergei was offering to provide Petrov with a strategy for bringing down the Gretzkys?'

'Is true,' said Arkady.

'And this actually happened?'

'Not exactly. As I say, Sergei impetuous man. He ask too many questions too quickly and Petrovs suspicious. They kill him.'

'So how do you know all this?'

'He call me. I take notes.'

We drove on in silence for several minutes while I tried to process everything I'd been told.

'One thing, though. You haven't told me what you were doing at the Institute that night.'

'Ah. Is for Sergei. Sergei ask me to break into Institute and find out everything about Gowers. He say very important. I promise Sergei I tell world.'

'Gowers?'

'Is true. Sorry.'

'So you knew the Vavasor papers weren't there at all?'

'Is also true.'

'And you let the rest of us get caught while you carried on looking?'

'Yes. I know. Sorry.'

'Did you find anything?'

'I have big – what is word? – dossier, yes.'

'Well, I guess that's something. But in that case, why aren't you back in the UK, talking to journalists?'

'I have one more thing to do here.'

'Right. And that is?'

'Later.'

'Hmmm,' I said. 'There are still a few things I don't understand,' I added.

'Go on,' said Arkady.

'Helen Matheson sent me to find Sergei.'

'Who is Helen Matheson?'

'That would take a long time to explain. She runs some kind of agency. Intelligence stuff.'

'And you work for her?' Arkady sounded incredulous.

'Yeah, scarcely believable, isn't it?'

'Sergei not mention Matheson.'

'He probably didn't even know about her. I wonder whose side she is on, though.'

'Tom, what did I tell you? No sides. Only chaos.'

'But is she a willing participant? Or is she just another tool being manipulated by some third party?'

'Perhaps you ask her?'

'Maybe I will one day.' There was one more question I hardly dared to ask. 'Arkady, if Sergei is dead, where are the Vavasor papers?'

'I was wondering when you ask that. Tom, I do not know. I sorry. Maybe the Gretzkys have them after all. Most likely the Petrovs have them and they lost in fire at hotel.'

So that was that, then. The papers were lost. I'd just wasted a month of my life chasing after a phantom. Not only that, but I'd almost got killed several times in the process.

'Is disappointing, I know,' said Arkady.

'It's more than fucking disappointing. It's a fucking disaster.'

'Why, Tom?'

'Never mind.' I sat in silence for a few more miles. The traffic was thinning out now. 'How did you find me?' I said eventually.

'I had message from Mikhail,' he said.

'Mikhail? Mikhail who?' I'd only met one Mikhail recently and it surely couldn't be him.

'Mikhail who drive you to Minsk. He old family friend.'

So it *was* that Mikhail.

'Good grief. So he told you about where I'd been and what I was doing?'

'Yes. He think you are mad. He knows you have papers from Alexei and that the Gretzkys are angry that you have

238

them. But when he hear about fire at hotel, he realises they even angrier than he think. He also hear rumour you escape.'

'So was that all about me?'

'Not you,' said Arkady. 'About papers. Alexei so angry he order strike to retrieve them. When strike fail, he so angry he destroy hotel.'

'But he's just a kid. Surely he doesn't have that much clout?'

'He oldest male Gretzky. Is patriarchal system.'

If I'd realised this, I probably wouldn't have stolen anything from him. It was barely an afterthought in the flurry of leaving the camp.

'That still doesn't explain how you found me,' I said.

'Ah, is simple. Mikhail ask father for details of calls made on his phone, I speak to them and with Mr Unsworth I find the pay dirt. Your father says hello, by the way. He interesting man. He try to sell me alpaca.'

'That would be him.'

'I no have space for alpaca. Is shame.'

'They're not his to sell, you know.'

'Maybe one day I have alpacas. I would like that.'

There wasn't any sensible way to respond to that, so I let it lie for a couple of minutes. Arkady turned off the main road onto a bumpy track. We continued on this way for a few miles.

'One last thing I don't get,' I said, 'is how did Alexei turn into some kind of weird mathematical genius? He's not exactly your typical crime boss, is he?'

'Alexei visit England when seven years old. When Gretzkys work with Vavasors. His father Bruno realise he clever kid and he bring him over to introduce him to the business. Alexei like Vavasor twins. They like him. They teach him lots of complicated mathematics. He learn very fast.'

'So that would explain how excited he was when we first met. He thought I'd actually worked with them.'

'Alexei is prodigy. He develop Vavasor algorithms. He also good chaotician. He is also—' At this point, Arkady turned into the front yard of a small farmhouse and came to a halt. 'He is also,' he continued, 'the way we bring Gretzkys down for good.'

'Really? How are you going to do that?'

'You have his precious papers?'

'Yes.' I patted my chest.

'Good,' he said. 'We arrange handover.'

Arkady opened the door and stepped out. I did likewise. A small flock of wildfowl was beginning to gather around our feet ready to escort us into the house. There was a heady scent of farm animals in the air.

'You're just going to hand the papers back to him?' I said as we walked on. 'Just like that?'

'Of course not,' said Arkady. 'We prepare trap. He admit papers are his. You hand papers over. He take papers. We spring trap.'

'But the papers are meaningless to anyone other than an expert. Honestly, I've tried to make sense of them. You'd never in a million years be able to establish that they were intended to do any harm to anyone.'

Arkady opened up the door and we entered a large kitchen mostly occupied by a vast circular table. Seated at the table was a middle-aged man, smoking a pipe and playing patience. A woman of a similar age was chopping vegetables. There was a pot on the stove and a smell of something good cooking.

'My parents,' said Arkady.

'Pleased to meet you,' I said. They both nodded politely in my direction.

'They not speak English,' explained Arkady.

'No need to apologise,' I said. 'It's not as if I speak any Belarusian. Although I intend to learn. I like your country very much.'

Arkady translated and they both smiled.

The father gestured towards a chair opposite him and said something in Belarusian that sounded like it probably meant 'Take a seat.'

'Take a seat,' said Arkady.

We both sat down.

'Does your mother want any help with the vegetables?' I said.

'Is fine. She enjoy preparing food.'

Judging by the frequency with which she had to stop to groan and stretch her back, I wondered how true this actually was, or if it was more the case that the patriarchy was still alive and kicking in these parts.

'So what do you plan on doing with Alexei's papers?' I said to Arkady. 'You'll need to find someone who can make sense of them if you're going to build any kind of case for the authorities here. I mean, I could suggest one or two, but they're all back in England.'

A door opened at the back of the kitchen and someone new walked in and my heart skipped a beat. Actually, it skipped several beats. It was the last person I had expected to see here, deep in the heart of the Belarus countryside, but someone who was none the less welcome for all that.

'You're wrong there,' said Dorothy, taking a seat next to Arkady's father. 'They're not all back in England, you know.'

Chapter 21

I gaped at Dorothy. It was, of course, wonderful to see her again, but I was keenly aware of the fact that she hadn't exactly run to my side of the table and flung her arms around me. In fact, she seemed largely indifferent to my presence altogether. She'd had her hair cut even shorter since I'd last seen her. It suited her.

'Hi,' I said. 'It's good to see you again.'

'Yeah, hi,' she said. 'I brought your passport.'

'What? Oh, right. Thanks. Thanks very much.'

'No problem.'

It wasn't so much that you could have cut the atmosphere with a knife. You could probably have cut it with a wooden spoon.

'Have you got Alexei's papers?' she said.

'Um, yes. Yes, I have. They're—' I awkwardly removed them from under the front of my shirt. '—here.'

I couldn't quite reach across the table to her so I ended up tossing them across in what must have looked like an unnecessarily dismissive gesture.

'Thanks,' said Dorothy. She began to pore over the pages of scribbled equations. After a while, she looked up.

'Are these in the right order?' she said.

'I… don't know,' I said. 'They're just how I grabbed them when I was leaving the Gretzky camp.'

'Right,' she said. 'So not necessarily, then.'

'Look, I'm sorry if that's going to cause a problem.'

'It's not a problem if I know. And I know now, thank you.'

She resumed her study.

I waited for a few minutes to pass, then I started to say something but ended up in an awkward fit of coughing. I tried again.

'I… I was hoping I could have got you the Vavasor papers,' I said.

'Well, never mind,' she said without looking up.

'No, I really did try. It's just that Sergei… I mean, I didn't manage to get to him before… I mean, he was probably dead long before I… well, sorry anyway.'

'Yes, well.' She still didn't look up.

'Look, about those WhatsApp—'

'Please, Tom,' she said, looking up at me now. 'Not now.'

'I mean, the thing is, it wasn't—'

'Seriously, Tom. This is not a good time.' Dorothy put her head down and resumed her trawl through Alexei's work.

'OK, OK,' I said. 'When would be a good time, then?'

Dorothy didn't dignify this with any kind of reply, and I had a strong impression that the appropriate time to discuss those WhatsApp messages would be a couple of millennia after the ice finally jammed shut the gates of hell.

I got up and walked outside. I sat down on the steps and watched the chickens doing their chicken stuff. It looked a decent kind of lifestyle, all things considered: there was plenty of food to be had, regular sex and the only thing to be worried about was making sure you were back in the coop in time to avoid the attentions of the local fox.

In the end, being human was just too complicated. There were too many expectations: your parents', society's and worst of all, your own. However much you tried, there was always the constant feeling that you weren't quite fulfilling all that promise you showed back in the day when you once won a runner-up prize for recorder playing in primary school.

And then there were those times when, despite all the odds, you accidentally managed to succeed in a league way above what you'd grown to think of as your own, and you began to dare to imagine that maybe this was a permanent promotion. There were also the dark days after each fall, when you realised that you'd been right all along. This was where you belonged, down in the depths with all the other losers. Welcome back to the last chance saloon, old friend, where were you? We missed you.

I felt a hand on my shoulder. I turned round and saw Arkady's muscled frame squatting down in the doorway behind me.

'It will take time, Tom,' he said.

'What will?' I said, before realising. 'Oh, I see. You think?'

'I think so. Drink?'

He held out a bottle of beer. I took it from him and took a mouthful. He sat down next to me and drank from his own.

'She want to come here, Tom,' he said.

'Maybe she fancies you,' I said, ruefully. I'd already lost one girlfriend to him, after all.

'No, Tom. She not like Lucy. She still want you, I think. But she hurt now. You have to fix her.'

'But I can only do that if she wants to be fixed. I'm not at all sure she does.'

'She does, Tom. Give it day or two, you have big shout at each other then all fine.'

'I wish I had your confidence,' I said. But in my heart of hearts the tiniest glimmer of hope was beginning to stir. Or perhaps it was just the beer that was making me feel better. Arkady stood up.

'Dinner nearly ready,' he said. 'I help with table.'

I began to stand up myself, but I felt a gentle pressure on my shoulder. 'Stay there, Tom. I call you when food ready.'

I got the impression that the last thing Arkady wanted when he was setting the table was me lurking awkwardly around in the kitchen.

It was my first hot meal for over a week and until the first mouthful entered my stomach, I hadn't realised quite how hungry I was.

'This is excellent,' I said. 'What is it?'

It turned out that whatever it was didn't really have a name. It was just some sort of stew that had been passed down from one generation of Arkady's family to the next. Whatever it was, it was nutritious and it also acted as a suitable cushion for the inevitable large amounts of drink that were served with it. This came in the form of Krambambulia, a formidable concoction of red wine and whatever spirits might happen to be lying around, mainly rum and vodka in this case, flavoured with honey and spices. When this ran out, we transferred our attentions to Arkady's father's home-made vodka, which seemed to have been put together using the exact same formula as Artem's and was equally powerful.

Inevitably, there was singing involved afterwards, although I did my best not to make too much of a fool of myself. Dorothy joined in as well, although as hard as I tried to catch her eye, she succeeded in avoiding me for most of the evening. Come midnight, we were all too tired to perform any longer and we all retired to bed. Dorothy and I were shown to separate rooms. I briefly considered the logistics of a sneak visit in the early hours, but there was no guarantee whatsoever of a favourable reception, especially as I hadn't showered for a week. As it turned out I was so tired I slept right through until ten o'clock the next morning anyway.

As I staggered into the kitchen for a late breakfast, I found Dorothy still at work deciphering Alexei's notes. There was no sign of either Arkady or his parents.

'Getting anywhere?' I said.

'Maybe,' said Dorothy.

'Sleep well?'

'Yeah.'

I stretched my arms and gave an exaggerated yawn. Dorothy ignored me.

'Look, I…' I began.

'What?' said Dorothy.

'No, nothing.'

'Oh, OK.'

'I mean. I don't know. Actually, I need something to eat. Is there anything around?'

'There's rye bread in the cupboard over there and cheese in the fridge. Teabags in the tin.'

'Cool. Thanks.'

I boiled the kettle and made myself a cup of tea. Then I cut a couple of slices of bread and added some of the cheese, before sitting down at the table opposite Dorothy.

'So what's it all about?' I said.

'You're speaking with your mouth full, Tom,' said Dorothy. 'We've spoken about this before, remember?'

'Yeah, OK,' I said, rolling my eyes as theatrically as I could manage but inwardly grinning from ear to ear. I finished eating and tried again. 'So?' I said.

'Well, it seems to be some kind of investment strategy on an absolutely massive scale. But it's all over the place. Look at this.' She moved round the table to sit next to me. 'See there? Twenty million into a stock here, thirty million into one over there. Couple of hours later, half of the first investment is yanked right out and chucked into the second one. Repeat this over the entire FTSE 100. Now normally that would be done through an intermediary so as to avoid messing up the price, but this note says that it should be done as loudly as possible.'

'To cause chaos,' I said.

'Exactly. Chaos. The entire purpose of all this is to cause a market crash. Maybe even bigger than 2008. You can bet that someone somewhere will also have a humongous number of put options set up which they can sell to everyone who gets caught.'

'How does that work?' I said. I never understood this stuff.

'OK, here goes,' said Dorothy. 'A put option gives you the right to sell a stock at a specified price at some point in the future, regardless of what the current value is. So if I were to buy a load of puts on Widget PLC at 99p per share when the price for Widget PLC happened to be 110p a share, you might think I was being a little over pessimistic and I'd probably get them at quite a discount. However, once you've driven that price down to 50p a share, you'll suddenly find, once you've covered any outstanding shares in Widgets that you happened to have left yourself with, a whole world of investors out there who will be suddenly very keen to hoover up whatever puts you happen to have left over.'

'So they can get at least 99p per share back on their investment rather then get stuck with 50p?'

'Exactly.'

'But how do these people get the market moving in the first place?'

'Chaos, Tom.'

'Ah. This is where I begin to get really lost.'

Dorothy took a deep breath. 'Well, I could try and explain it if you want.'

I was in two minds about this. On the one hand, it was good to be having something close to a normal conversation with Dorothy, and the longer it went on, the closer we might be getting to a reconciliation. On the other hand, there was every chance I would end up with my head spinning fast enough to play a halfway decent tune if you dropped a stylus on it.

'Go on, then,' I said.

'Are you sure?'

'Oh, what's the worst that can happen?'

'Your brain might explode.'

'It's only a brain.'

'Fair enough,' said Dorothy, grabbing a blank piece of paper to write on. 'OK, chaos theory is all about the ways in which real world systems evolve over time and whether they become stable or unstable.'

'Cool,' I said. This seemed OK.

'So what do we mean by real world systems? Well, we usually start off with rabbits.'

'Rabbits. Right.'

'Specifically the ones in the garden of a bloke called Leonardo of Pisa.'

'Da Vinci?'

'Different Leonardo. Anyway, he noticed this progression, which went something like this…' Here, she scribbled a formula on the piece of paper.

$$X_{n+1} = X_n + X_{n-1}$$

'…where X_n is the number of breeding pairs of rabbits in one iteration of the population. So if, for example, we start with X_0 and X_1 both as 1, we end up with the following sequence…' She added the following sequence of numbers underneath her formula:

1, 1, 2, 3, 5, 8, 13, 21…

'Recognise that?' she said, looking at me hopefully.

'Um…' I said, desperately trying to remember. It did look familiar.

'Maybe if I tell you that Leonardo of Pisa was more commonly known as Fibonacci.'

'Ah, that.'

'Yes, that. Basically, the number of breeding pairs of rabbits goes up in a Fibonacci sequence.'

'Right. And that's going to look pretty chaotic after a while? I mean, you'll get hundreds of rabbits running around everywhere. Which is definitely chaotic. So that's all this is about?' This was one of those rare occasions when I felt as if I was understanding everything.

'Well, yes,' said Dorothy. 'Sort of. Except it's not chaotic in a particularly interesting way.'

'Oh,' I said. Evidently it wasn't that simple.

'The thing is,' continued Dorothy, 'in the real world, you have constraints on the rabbit population. They run out of food and breeding slows down. So it all gets a bit more complicated and you need a different type of equation altogether.' While she'd been talking, she'd started doing that thing where she was spinning her pen around in her hand at high speed. She did this every time she got excited about something. I'd never seen anyone do it until I met her but it turned out that Ali was a dab hand at it too, so I assumed it was a developer thing. I tried it once or twice just to feel I was part of the team, but the pen just flew off in a random direction and I was told in no uncertain terms to give it a rest before I broke something.

'OK,' I said. 'Go on.'

'Right. So let's simplify things a bit first. In our new simplified model, let's say that the rabbit population goes up by a factor of r every year. So we get something like this.' She stopped spinning her pen at the exact point where it was aligned with the paper and wrote this:

$$X_{n+1} = rX_n$$

'OK?'

'I… think so.'

'But the trouble with that is that we end up with an exponentially increasing population. If r was two, for example and the value of X in the first year was ten, we'd end up with twenty in year two, forty in year three, eighty in year four and so on.'

'Right.' This was making some kind of sense, but I had an uneasy feeling it was all going to dive off a cliff very soon.

'The thing is, it doesn't take account of limited resources, so we need to put in some kind of damping factor. So let's reformulate things.' Here she modified her equation.

$$X_{n+1} = rX_n (1 - X_n)$$

'X_n now represents the current population, expressed as a fraction of the maximum possible. OK?' she said.

'Not really,' I said. 'In fact, not at all. Why do you have to do it like this?'

'The thing is, if we have that new one minus X term, every time the population approaches its maximum, X gets close to one and that bit gets close to zero, so next year's population dips.'

'Ah… hmmm, maybe.' I was teetering on the edge of the cliff, but still clinging on by my fingertips. 'Go on, then.'

'Right. Well, what happens with this model is that for values of r less than one, whatever value you start with, the population eventually becomes extinct. For r between one and three, you end up with a stable population. But with r between three and around three and a half, the population bounces between two fixed values. One year it's up, the next year it's down. When r gets a bit bigger, you get another bifurcation and it splits again, so the population now switches between four different values instead.'

'Another what?'

'Look, I'll show you.' She drew a graph with a single line increasing in a curve and then splitting into two, then four, then eight and so on, until she got to a point where she just started scribbling.

'What's happened there?' I said.

'When r gets to around 3.56995, we get chaos. The population can take any value from one year to the next. There's no pattern. All bets are off.'

'Oh.'

'Yeah, but what's really weird is that if you push on through, you'll find patches of stability beyond the chaos. At certain values of r, you might find the population bouncing between three different numbers. Or six. Or anything in fact. Islands amid the turbulence. And the thing is, despite it being chaotic, it's all predictable if you know what r is. Which is what makes it interesting to apply to the financial system. If you can create apparent chaos while at the same time being able to predict how it's going to play out, you can make a lot of money. The problem, though, is that you're dealing with a system that's about a thousand times more complicated than a bunch of rabbits.'

'Bloody hell.'

There was a period of silence between us.

'I've missed you,' I said.

'Yeah, well,' said Dorothy.

I tentatively put my hand on hers. Then the front door opened and Arkady walked in. Dorothy withdrew her hand.

'Tom!' he said. 'You're up. You sleep well?'

'Very well, thank you. I feel much better today.'

'Is good. And Dorothy? You make the progress?'

'Yes,' she said. 'I was just saying to Tom, they're planning some serious market manipulation here. But they're going to need so much capital to do it, I can't begin to imagine where they're going to get the funding from. Surely even with all the dark money they have from whatever scams they're up to, they won't have enough to do something on this sort of scale?'

'Is Gowers,' said Arkady.

'But I thought he was after the papers for himself?' I said. 'So he could set up his own chaos operation?'

'This is number one choice. Number two choice is find clever kid in Belarus to do work. Maybe Petrov. Maybe Gretzky. Maybe both. Is also good because is foreign country.'

'Oh, come on, Arkady. They can't be involved with both the Gretzkys and the Petrovs. Last time I saw that lot together in the same place they were shooting at each other. They aren't exactly big mates, Arkady.'

'Tom, Tom,' said Arkady. 'You forget again. Is not about sides. Sides not matter any more. Is about chaos.'

'But why Belarus?'

'Why not? My country good place for Institute to do business. Is young country. My people still innocent about capitalism. And is easy to hide here. No one has heard of us.'

'We've heard of you,' I said. I felt bad.

'I know, Tom,' said Arkady, giving me a friendly pat on the shoulder. 'But Minsk tourist industry is not big. Maybe one day, Tom. You and me, we set up tour company, eh?'

I looked up at him. I really wasn't sure if he was joking. There were worse things I could end up doing.

'So,' said Arkady to Dorothy. 'How much longer you need?'

'I've nearly finished,' she said. 'Might need to run a few things past Patrice, though. She's quite hot on this area.'

'But you finish today, yes?'

'I think so. I'd still like to know what you're planning, though.'

'Oh,' I said. 'So you don't know either?'

Arkady smiled. 'Only I know, Tom. You have to trust me.'

'Do we have a choice?'

'Not really.'

For a moment, Arkady seemed lost in thought. Then he snapped out of it, turned and went out again without saying another word. He seemed uncharacteristically on edge this morning. Most of the time, he was a stolid, muscular zone of stability around which the rest of the world nervously flitted. But today he seemed a different person. He was twitchy. And a twitchy Arkady was a dangerous Arkady.

'Are you as worried as I am?' I said to Dorothy.

'I'm sure Arkady knows what he's doing,' she said, without looking up from Alexei's work.

'Yes, but is it safe?'

'Unlikely.'

'Great.'

Dorothy didn't say anything more, so I went back upstairs and took a much-needed shower. When I came back down, she was on the phone to someone, who I took to be Patrice. I lurked in the kitchen doorway, eavesdropping on what Dorothy was saying.

'So you're saying he's got that completely wrong... so it should be beta cubed instead of squared... yes, I thought that too... I know, it would still be bad... OK... no, please don't tell me anything about the snakes...' Patrice kept reptiles. Lots of them. Every so often, one of them would escape, which didn't go down well with Dorothy, who tended to hide in the kitchen when this happened. 'Ali OK?... yes, Tom's here too...' My ears pricked up at this point. '...no, I've no idea either... might take a while... nope, I don't know either...'

Well, that could have meant anything. I went up to my room and lay on the bed for a while. It seemed to me that the only way I was ever going to prove anything to Dorothy about my relationship, or rather my lack of a relationship, with Helen Matheson was for us both to confront her. As soon as we got back to the UK, we would go to Margate and sort it out once and for all. Until then I would have to bide my time.

I got up and looked out of the window. The farm itself was little more than a smallholding. Apart from the chickens, Arkady's parents had a few sheep and a number of rangy goats, all crammed together in a couple of acres. Beyond that, the fields belonging to the old collective farm next door stretched out to the horizon, criss-crossed by dirt tracks. The only thing interrupting the view was the occasional barn.

After a while I heard voices downstairs and I decided to go down and join them. Arkady and his parents were gathered in the kitchen and Dorothy was explaining to Arkady what she'd found in Alexei's papers.

'The thing is,' she was saying. 'He's got it almost right, but there's clearly still quite a lot of the theory he doesn't understand properly.'

'So does that invalidate everything?' I said, butting in.

Dorothy frowned at me. 'Not entirely,' she said. 'They could still cause a lot of damage.'

'But you can make sabotage?' said Arkady.

'I think so. I'm pretty sure I can make a few alterations that Alexei won't notice.'

'Ah, so that's the plan,' I said. 'Neat.'

'Yes, is neat,' said Arkady.

'Will it really work?'

'We must hope so.'

I still had a feeling I wasn't being told the whole story, though.

'OK,' said Arkady. 'I make call.'

'Hang on,' I said. 'So let me get this straight. You're offering the Gretzkys their calculations back, only subtly adjusted so they don't bring the world's economic system to its knees, right?'

'Yes.'

'So what are you asking for in return?'

'Money,' said Arkady, pointing towards the ceiling. 'Roof needs repairs.'

'How much money?'

He shrugged. 'Ten million rouble. Is fair.'

'Jesus.'

'So I make call, OK?'

'Well, yes, go ahead,' I said. 'Go right ahead.'

'Also I make one other call,' said Arkady.

'Who to?' said Dorothy.

'Gowers.'

Dorothy and I both looked at him in alarm.

'What?' I said. 'But… that doesn't make any sense at all. Why would you call him?'

'I make him offer too.'

'But you can't offer the Gretzky papers to two different people,' I said.

'Is true. I offer him Vavasor papers.'

Chapter 22

Dorothy was genuinely shocked.

'But you can't do that!' she said. 'First of all, you don't have them, and secondly, even if you did, there's no way I would even think of letting you give them away to anyone else, least of all Julian Gowers.'

'Maybe we pretend,' said Arkady. 'Maybe I tell lie. At least this way we find out for sure he not have them already.'

'But why?' I said. 'Are you trying to get money out of the Institute too?'

'He offer me twenty million rouble.'

'That's a lot of roofing, Arkady,' I said.

Arkady shrugged. None of this was making any sense at all. Up until this point, I'd assumed that Arkady had a solid, workable plan. Unfortunately, I now realised that Dorothy and I were in the hands of a lunatic. There wasn't going to be much point in getting back together with her now, as we were unlikely to survive the next twenty-four hours.

'Arkady,' I said as firmly as I could. 'This is a dangerous, crazy idea.'

'Is dangerous, yes,' said Arkady. 'Crazy, no.' He turned and spoke to his parents in Belarusian and they both nodded in agreement. 'See? My parents like idea too.'

'Well, I can see where you get your insanity from,' I said. 'No offence.'

Arkady laughed, a little too wildly for my liking. 'Anyway,' he said. 'Is decided.'

'Hang on, don't we get a say in this?' I said.

'Yes,' said Dorothy. 'What about us?'

'Sorry,' said Arkady. 'But no. This my plan now.'

With that, he took out his mobile phone and left the room to make the calls.

'Well,' I said. But Dorothy was hard at work again now, making her changes to Alexei's algorithms. I wondered what the point of all this was any more. I turned to Arkady's parents, who were engaged in getting lunch ready. They didn't seem the slightest bit bothered.

The rest of the day passed in something of a blur. My general sense of rising panic wasn't helped by the fact that after lunch I was enlisted to help with fortifying the front of the house with sandbags and covering up the downstairs windows with large slabs of chipboard.

'Is precaution,' said Arkady. This didn't really help.

Meanwhile, Dorothy was engaged in making sure her changes would ensure that the Gretzkys and the Institute were only likely to inflict flesh wounds on the world financial markets. This was the only part of the plan that made any real sense, and even that was pretty sketchy. To be perfectly honest, I found it very hard to believe that Alexei was going to fall for it, but Arkady seemed very keen to go ahead with the idea. The more likely outcome to my mind was that Alexei would take one look at it, decide that it had been tampered with and order his henchmen to kill us on the spot. The sandbags were not going to save us.

Dinner that night was a distinctly less raucous affair. No one said much, either in English or Belarusian, and the drinking was restrained. At around ten o'clock, Arkady yawned and announced that he was turning in, advising the rest of us to do likewise. We would need to be as alert as possible tomorrow. Who knew what was really going to happen when the rubber of the plan met the road of real life?

If the previous evening's meal had been subdued, breakfast was positively funereal. When I accidentally brushed against Arkady's father pouring myself a cup of tea and said 'Sorry', the entire room turned and stared at me as if I was some kind of maniac. It was like the Dawn of the Trappists.

Our visitors were scheduled to arrive at eleven o'clock, so we made sure the chickens were all herded out of the way by half past ten. We didn't want any stray hens wandering into the middle of the negotiations. A trestle table was set up to one side of the front door, laden with bottles of vodka and shot glasses. Arkady, meanwhile, was tinkering around with a load of cables in the courtyard in front of the house. When I asked him what he was up to, he just tapped his nose and told me to wait and see. I wondered if there were explosives involved, although I couldn't see any.

Finally, at five minutes to eleven, a small cloud of dust appeared on the horizon, heading our way, and before long a silver Honda Civic came into view.

'I insist he come alone,' said Arkady as the visitor drew up outside the house. Arkady stepped forward to greet him and Dorothy and myself took up positions on either side.

The car door opened, and out stepped a balding, early middle-aged male, dressed down as if for a geeks and freaks fancy dress party and not really convincing anyone. He eyed us up, trying to establish who was a threat to him and who wasn't, and then attempted a smile but only achieved a kind of modified sneer.

'I thought it was going to be just between the two of us,' said Julian Gowers.

'These my associates,' said Arkady. 'Tom?' He nudged me forward. I wasn't sure what he was indicating by this, so I raised a querying eyebrow. He responded by waggling his hands vertically up and down. I took this as my cue to step forward and frisk the new arrival, who intensified his sneer as I checked him out for concealed weapons.

'Do I know you?' he said.

'I don't think so,' I said.

'You look familiar.'

'I've got one of those faces.' I turned to Arkady. 'All clear,' I said.

'I'm sure I've seen you somewhere before,' insisted Gowers.

So maybe they did have a file on me back at the Institute after all, I thought to myself.

'And you,' he said to Dorothy. 'I've seen you before, too.'

'Maybe last year's "Women in Software" conference?' she said.

'I doubt it.'

'No, that one probably wouldn't have registered with you.'

'Probably not. Look, Morozov, can we get this over with?'

'Is no rush,' said Arkady. 'You like drink?' Here he indicated the table next to us.

'Oh, let's just cut the crap and get on with it.' There was a strong sense of suppressed anger emanating from Julian Gowers, and I wondered if Arkady realised this and was deliberately stoking the flames.

'Sure,' said Arkady. 'But is the Belarusian way.'

'I do business the English way,' said Gowers, coldly.

'And that is?'

'I give you the cash, you give me the papers, I fuck off home. So can we just—'

'Is fine. So tell me, why Vavasor papers so important to Institute?'

'You know full well why.'

'I make sure before I hand over.'

'Well, then. The Vavasor papers contain some of the most advanced investigations into certain fields of mathematics. A lot of people would kill to get their hands on them.'

'Including yourself?' said Arkady. I glanced at Dorothy. She remained impassive.

'You would find that difficult to prove,' said Gowers. 'Several have tried, but none of them has ever been successful. I would warn you against making idle threats, Morozov.'

'You know my friend Sergei?'

I glanced across at Dorothy. This was clearly as much news to her as it was to me.

Julian Gowers looked alarmed, too. 'I'm not happy with the tone of this conversation,' he said, frowning. 'Can we just get on with—'

'Has Sergei ever worked for you?'

I looked at Dorothy again. Arkady was definitely striking out on his own now. I wondered what other revelations were about to drop.

'I am under no obligation to answer any of these questions.'

'Has Sergei ever worked for you, Mister Gowers? This important.'

'Lots of people have worked for me, Morozov,' said Gowers. 'I don't keep a mental record of every single fucking one of them.'

'Did you kill Sergei?'

'No. But let me just say that he was making a bloody awful nuisance of himself, so I wasn't that surprised when I heard he was no longer with us.'

'How do you know he's dead?' I said. 'No one else does, apart from us.'

'Oh, piss off, Saga Norén,' said Gowers. 'Word gets around is all I'm going to say. Apart from adding that if you people persist with this line of enquiry, you may also find that I make your lives singularly difficult.'

'I not afraid of you, Mister Gowers,' said Arkady.

'And if you insist on continuing with this unnecessary bullshit, the deal's off.'

Arkady threw his hands up as if he didn't care either one way or the other. 'Is fine. I believe your need greater than mine.'

'But twenty million roubles is a lot of money,' said Gowers.

'Not for you. Also, money not everything.'

Julian Gowers looked around the courtyard. A couple of chickens wandered in through the open gate. 'Look at this place,' he said with contempt.

'Is home,' said Arkady.

'Is home, indeed. Well, let's get this shitty charade over with, then. I'm beginning to feel contaminated just by being close to this revolting hovel.'

Arkady shook his head in disgust but said nothing.

'Well?' said Gowers. 'What are we waiting for? Where are the papers?'

Arkady shrugged. 'I no have them,' he said, eventually.

Julian Gowers frowned. 'I beg your pardon?' he said.

'I change plans. Sorry.'

The atmosphere in the courtyard was developing a chilly edge.

'What do you mean?' said Gowers, taking a step forwards. Ice crystals were beginning to form in the air.

'I am – how you say? – impetuous,' said Arkady. 'I decide I no like this plan so I make new one.'

'Mr Morozov, I hope you are not implying that I have made a wasted journey to this godforsaken shithole at the edge of nowhere. You made explicit assurances that you had the Vavasor papers in your possession.'

I noticed that Arkady was surreptitiously looking at his watch. He was stalling, although Gowers was too angry to notice. Surely the Gretzkys should be here by now?

'Oh, I have Vavasor papers,' he said, airily. 'But perhaps I find better offer.' He was playing with him now.

'This is absurd,' said Gowers. 'I won't be messed about, Morozov.' He turned to go back to his car but Arkady called after him.

'One moment,' he said.

'What's going on?' said Gowers.

Arkady cocked an ear towards the track leading to the farmhouse. 'This,' he said.

A minute later, a Jeep drew up behind Gowers's hire car and Alexei Gretzky hopped out, followed by Anya and Novikov. I noticed that Anya and Novikov were both armed, and I wondered if Arkady had factored this into his calculations. Part of me was also wondering if this might be a good time to disappear back into the house, but the other part of me was desperate to see how this turned out. I sensed a similar conflict going on in Dorothy, and in both our cases the second option won. If popcorn had been on offer, we'd have snaffled up two portions each, one with salt for the main course and one with caramel sauce for dessert.

Meanwhile, Julian Gowers was describing a delicate circle as he sidled his way round to the opposite side of the courtyard from the new arrivals. Arkady nudged me forwards again to frisk them.

'Is OK,' he said to me. 'Go on.'

I wasn't keen on the idea at all, but I felt I had to do something for appearance purposes.

'Guns, please,' I said to Anya and Novikov. They both laughed and shook their heads. Then Alexei turned to them.

'Go on, guys,' he said. 'These people are cool.'

With great reluctance, the two of them handed over their weapons and I carried them gingerly over to the trestle table. Knowing my luck, there was every chance that one or both of them would go off in my hands.

Having sorted out the domestic arrangements, Alexei and his two companions looked at Arkady, then at Dorothy and myself and finally at Gowers.

'Why are you here?' he said to Gowers.

'That, my friend, is a fucking good question,' he said.

'That is not an answer.'

'Perhaps Mr Morozov has a big plan for all of us,' said Alexei, raising an eyebrow in Arkady's direction. Arkady said nothing.

'Hey, Rory,' he said, turning to me. 'Or should I say not-Rory? Good to see you're still alive.'

'No thanks to your efforts,' I said.

'Fair comment. But you stole my work.'

'You could have asked for it back.'

'Would you have given it to me?'

'Probably not,' I said.

'Well.'

'But maybe we'll sort it all out today.'

'Or maybe I order my friends here to take back their guns and shoot everyone. Apart from our friend Mr Gowers over there. We need his money.' Alexei raised a hand to him in greeting. Gowers gave the slightest of movements in return. He looked embarrassed to be here at the same time as them, and I could easily understand why. For there it was, right in front of us: the proof of collusion between the Institute for Progress and Development and the Belarusian mafia.

Arkady stepped forward now, his hands raised in truce. 'Friends, friends,' he said. 'Let us not make the threat. We do good deal. We all happy. We go home.'

What was he planning? I still didn't see how it all fitted together. There was no possible deal he could do that would leave everyone happy.

'Who is this woman?' said Alexei, pointing at Dorothy.

'This Dorothy,' said Arkady. 'She Vavasor expert.'

'Is she for real?'

'I am,' said Dorothy, 'and if you'd like me to explain all the errors I've corrected in your work, I'd be very happy to.'

'Ha!' said Alexei. 'Very funny. Who are you really?'

'I am who I say I am. One moment.' She nodded to Arkady, who unexpectedly took his phone out of his pocket. He pressed it a few times and the whitewashed wall on the opposite side of the courtyard to the house blazed with light. He pressed another button and an extract from Alexei's calculations appeared in sharp, blown-up detail. One section had been ringed in red, with some changes made to it in Dorothy's handwriting. Alterations in plain sight. Very neat.

Alexei stared at it and then started laughing.

'Very good. Very good indeed,' he said. 'Maybe you would like to come and work for us?'

'Maybe we talk later,' said Dorothy.

'Yes, maybe we will,' said Alexei.

Arkady looked at everyone in turn, staring hard into people's faces and nodding fiercely at them. Then he said, 'Well, ladies and gentlemen, I thank you all for coming here. Today will be good day for Belarus. But first, I must ask to watch short film.'

'Oh, for Christ's sake,' said Gowers. 'You've already kept me here far too long.'

'This true,' said Arkady. 'We must finish this now. But we must watch film first. Alexei?'

'I'm good,' said Alexei. 'I like movies.'

'Good,' said Arkady. He pressed his phone a few more times and the extract from Alexei's code was replaced by a picture of a man, speaking. However, no sound was coming from his lips. Arkady swore and then pressed his phone a few more times until a voice began to emerge from speakers high up in the rafters above us. Then he wound back the video to the beginning and I realised why the voice seemed oddly familiar. I'd heard that voice back on the night when he'd delivered a case to me laden with explosives.

'Hello,' said the voice. 'My name is Sergei Kravchenko. I was the man who prepared the bomb that killed four members of the Gretzky family. If you are watching this recording, it is very likely that I am already dead.'

Chapter 23

'Bastard!' said Alexei. 'Is he really dead, or can we still kill him? I want to kill him. Slowly. He killed my father.'

Arkady paused the film. 'He dead,' he said. 'But he worth listen to.'

'Bastard,' muttered Alexei again. 'He friend of yours?'

'He was friend, yes.'

'Maybe I kill you.'

Arkady spread his arms wide. 'Go ahead.'

Alexei shook his head. 'Go on,' he said. 'Go on.'

The film continued. 'I speak in English,' said Sergei, 'because although the Belarusian people here speak English, English people are terrible at Belarusian. Is the same every time.' His accent was barely discernible.

'What's the point of this?' muttered Julian Gowers. Arkady sighed and paused the film again. 'We have a transaction to complete and then I have a plane to catch,' added Gowers.

'Please watch,' said Arkady, pressing a button to continue.

'I would like to tell you a little about what took place that night,' continued Sergei. 'As you may be aware, Alexei, your father and his associates had kidnapped a young woman in order to extort something from her friend. If my friend Arkady has been successful, that young woman, Dorothy Chan, is here with us all today, along with her friend Tom Winscombe.'

'No!' said Alexei.

'It's true,' said Dorothy.

'Every word,' I added.

'The object that was to be swapped for Dorothy was a briefcase containing the mathematical papers of the Vavasor twins. Arkady came to me to ask for help in executing the transfer, but what I didn't tell him was that I had been planning for some time to take revenge on Alexei's father Bruno for what he did to my brother Maxim. So I removed the papers, packed the briefcase with explosives and returned it to Arkady to pass on to Tom to exchange for Dorothy. I changed the combination to make sure that Tom didn't accidentally blow himself up on the way to the meeting.'

'Cheers,' I said under my breath.

'Bastard,' said Alexei.

'Much to my surprise,' continued Sergei, 'the plan worked perfectly. Dorothy was freed and the Gretzkys were killed. I am sorry, Alexei, but you need to know that your father was a very bad man.'

'Bastard,' said Alexei again.

'However, the question that no one asked,' said Sergei, 'is how did Sergei get hold of a consignment of C4 high explosives? Interesting, is it not? Sergei Kravchenko, security expert for Isaac Vavasor, good, law-abiding guy, right? What connections does he have?'

He gave a dramatic pause, as if waiting for someone to answer his question. No one did.

'The answer is that Sergei was given them by another organisation who also wanted Bruno Gretzky dead.'

'What?!' I said. I turned to Dorothy and I saw the shock on her face. It was one thing for us to have got caught up in Sergei's personal vendetta, but quite another for it to turn out to be arranged by some other bunch. Alexei seemed less surprised than we were. He was probably used to the idea of people wanting to kill his family. Julian Gowers was stony-faced.

'They offered me very good money,' said Sergei, 'and I was very happy to accept. Their objectives and mine were strongly aligned, although their reasons were more commercial. Bruno Gretzky was a difficult man to deal with, and perhaps Alexei would be an easier prospect.'

'Bastard,' said Alexei yet again. 'Who is this?'

'I'll come on to who they are later,' said Sergei in the video, as if predicting Alexei's question. 'For now, consider my position. I should perhaps have gone back to Isaac Vavasor, returned his brothers' papers to him and then found somewhere to hide for a while. But before I could do this, I found out that the organisation that paid me to kill Bruno Gretzky now knew that I had the papers. And they wanted them from me.'

'Who are these people?' said Alexei. 'Tell me now and I kill them!'

'So I left the country immediately and came to Minsk,' said Sergei. 'At first, everything is fine. I am safe. The papers are safe. But then I get lonely and I call my Carla and she comes out to see me. She is followed to Minsk and now they know where I am. One night, we go out to a film and when we come home, the flat is a mess and the papers are gone.'

'Are you following this?' I whispered to Dorothy.

'Shut up, Tom,' said Dorothy.

'Do we have to endure any more of this nonsense?' said Gowers.

'Five minutes,' said Arkady, holding up a splayed hand.

'I hear a rumour,' continued Sergei, 'that this time the Petrovs are responsible. The rumour also says that they have no one to interpret the papers, which gives me an idea: I offer myself to the Petrovs as their mathematician. That way I can steal papers back. Maybe this time I try somewhere else to live. At first, I think this plan is working, although they are reluctant to let me see the papers and now I think I have found out what the problem is. It turns out that the people who paid me to kill Bruno Gretzky and who also want the Vavasor papers have identified me to the Petrovs. I am now hiding in the Petrovs' hotel but it will not be long before they find me and that will be that.'

Sergei's voice had increased in speed and pitch as he spoke this last part and he was clearly in a state of some distress now.

'I think they are coming for me now, so I will stop filming and transmit this. But you must know now that the man who arranged the delivery of explosives to kill your father, Alexei, is also present today. He works for the Institute for Progress and Development but you know him as Julian Gowers.'

The video came to an abrupt stop at this point, and then real life took over, in a kind of weird slow motion, as Alexei lurched forward with a cry of 'Fucking bastard!', evading the clutches of Anya and Novikov, pushing the table towards Arkady, who fell back against the front of the house. As the table tipped over, Alexei grabbed Anya's semi-automatic rifle off it, staggering backwards while loosing off a burst that caught Gowers in the face and chest.

Before any of us in the courtyard managed to regain any idea of what was going on, another single shot rang out from above us, and the gun flew out of Alexei's hand. We looked up and saw Arkady's mother, grim-faced, staring back at us down the barrel of a shotgun. It seemed that Artem's mother was not the only babushka with sharpshooting skills.

Meanwhile, Arkady had recovered his composure and, holding Novikov's rifle in one hand, strode forward and collected Anya's, while Alexei sucked his bloodied hand, whimpering.

'Now what?' I said.

'One moment,' said Arkady, going over to check on the state of Julian Gowers's body. He bent down, took one look and then stood up again, shaking his head. He went into the house and then re-emerged shortly afterwards, followed by his mother carrying a sheet, which she proceeded to lay over the corpse.

'You let him do that,' said Dorothy. 'Didn't you?'

Arkady shrugged. 'War is not fair,' he said. 'And this is war.'

'You still shouldn't have let it happen.'

Arkady ignored her.

'But what happens now?' I said. Alexei was out of action for now, but Anya and Novikov were looking decidedly twitchy.

The answer came in the form of the arrival of another couple of vehicles. The first one was a white van, from which emerged half a dozen heavily armed policemen. They quickly took control of the scene, binding the three Gretzky gang members' wrists with plastic ties, paying little attention to Alexei's cries of pain as they did so.

The second vehicle was a small red Suzuki 4x4. The door opened, and a familiar figure stepped out, pushing his sunglasses onto his forehead. Mikhail came over and gave Arkady a warm embrace.

'Good work,' he said. Then he turned to me and gave me a big embrace as well. 'You did well, too, my friend. You are definitely mad though.'

Dorothy stepped forward and shook Mikhail's hand.

'So you are Dorothy, then?' he said. 'Tom told me so much about you. I am glad you were here to see the end of the Gretzkys and the Institute.'

'But is it really?' she said.

Arkady pointed up at the eaves of the house, where I could just make out a small blue light flickering.

'Is all recorded,' he said. 'Alexei guilty of murder. Institute guilty of mafia. If not the end, they are definitely badly injured.'

'I still have some friends left in the police,' said Mikhail. 'We finish this now.' The leader of the police squad came up to him and said something to him in Belarusian.

Mikhail tapped him on the shoulder and nodded. Then the policemen got back in the van and drove off.

'They send ambulance for body,' said Mikhail.

'Come into the house,' said Arkady. 'We need drink.'

The six of us sat round the kitchen table, each with a large glass in front of us, which Arkady proceeded to fill to the brim. It wasn't even lunchtime, but in the circumstances, it was late enough in the day to start drinking. We toasted each other, our respective countries and an end to the shadowy organisations that were wrecking our world. Then we all stood up, sang – or at least, in Dorothy's and my cases, mimed along to – the Belarusian national anthem and drained our glasses. Arkady topped them up to the brim again.

'Arkady and I grew up together,' explained Mikhail, after we had all toasted each other, the future of Belarus and the death of the mafia. 'When Chernobyl happened, his folk—' here, he indicated Arkady's parents '—moved out, but mine decided to stay. My father says hi, by the way.'

'And Grandma?'

'She does not say much.'

'Well, no change there,' I said.

'Maybe she was just scared of you, Tom.'

'Me? There's nothing to be scared of.' I looked round at the others seated at the table. They seemed to find this as amusing as I did.

'I don't know,' said Mikhail. 'You have an interesting track record, my friend. In the last few weeks, you have provoked a confrontation that resulted in the destruction of the Petrov family, helped to engineer the downfall of the Gretzky family and inflicted considerable damage on the Institute for Progress and Development.'

'I wasn't exactly trying.'

'But you seem to be the perfect agent of chaos.'

'Speaking of which,' said Dorothy. 'What do we do with Alexei's papers?'

'Hand them over to the police?' I said.

'What for?' said Arkady. 'The police not understand the mathematics.'

'No, but surely they're evidence for something,' I said.

'For what?' said Mikhail. 'It's far too complicated for any kind of legal case. Take it from me. I know about these things. But tell me, Dorothy, is any of this useful to anyone?'

Dorothy thought long and hard about this. 'I don't think so,' she said. 'Unless they want to try to wreck the markets.'

'But you fix it so it won't,' said Arkady.

'Actually,' said Dorothy. 'That's what's bothering me. Why did you bring me here to water down the calculations if your plan didn't actually involve handing them over?'

'It was backup plan,' said Arkady. 'In case Mikhail not persuade police. Also, she was in the script for Sergei's video.'

'Great,' said Dorothy.

'Hey, don't knock it,' I said. 'At least you were part of a plan. I've just been winging it.'

Dorothy ignored me.

'So what are we going to do with them?' she said.

'Burn them,' said Mikhail.

Dorothy looked shocked. 'But you can't just do that,' she said. 'What if—?'

'What if what?' said Mikhail. 'You said they were no use to anyone.'

'But there's some of this that's quite new and interesting, even if Alexei's made mistakes.'

'Are you willing to risk someone else using it?'

'You can't just go round burning things.'

'I think we can,' said Mikhail, jerking his head towards the corner of the kitchen, where there was a large cast-iron stove that was used to heat the water up. 'Arkady,' he said, getting up, 'do you have the papers?'

'Mikhail is right,' he said to Dorothy. 'Is best to burn them.' He handed the documents over to Mikhail, who strode over to the corner, opened up the front compartment and threw them in.

'There,' he said.

I suddenly remembered that I still had Dr Milford's papers too. If everyone was right about burning Alexei's calculations, then they should be destroyed as well. I went and fetched them from where I'd left them in my room, remembering just in time to remove the one on the back that I'd scribbled the address of Helen Matheson's safe house on. I stuffed that in my pocket and walked over to the stove, the remainder in my hand.

'What are you doing, Tom?' said Dorothy.

I bent down and began feeding the papers into the flames.

'Just getting rid of this lot as well,' I said.

'What are they?'

'Oh, just some papers that I nicked from the real Rory Milford.'

'Who?'

I pushed another batch of papers into the stove.

'The mathematician bloke I was supposed to be pretending to be. Did I forget to mention I bumped into him at the Petrovs' hotel?'

'Yes, you did.' Dorothy was looking anxious. She had pushed back her chair and was preparing to stand up.

'What's wrong?' I said, stuffing the rest of the papers into the fire.

Dorothy suddenly hurtled over to where I was crouched down. 'Give them to me,' she said.

'What?'

'The papers. The ones you haven't burnt yet.'

I spread my hands wide. 'There aren't any left,' I said.

'What? None?'

'Dorothy, what is this all about? You've gone really weird.'

'Tom,' she said. 'Please tell me you haven't just burnt all Dr Milford's papers.'

'Um, that would be difficult,' I said. 'Because I think I have.' Then I had another thought. 'Oh, hang on,' I said, reaching into my pocket. 'There's this I suppose.'

I handed the single sheet over to Dorothy, showing her where I'd written down Helen Matheson's address. But she wasn't interested in that at all. Instead, she turned the piece of paper over and began to read the other side. Her reaction to it was unexpected. She dropped to her knees and let out a despairing wail.

'What's wrong?' I said. 'Are you all right?'

Dorothy seemed to be struggling for breath, and it was all she could do to gasp out a few words.

'Tom', she said. 'Look! The glühwein stain! You've only gone and destroyed the Vavasor papers.'

Chapter 24

Once upon a time, there were twin brothers, Archimedes and Pythagoras Vavasor. They were both absurdly intelligent and they both won scholarships to read mathematics at Cambridge University at the age of fifteen. Unusually in an age of specialisation, their interests weren't confined to one particular field and they would from time to time decide, for example, that fractal magnetohydrodynamics might be a worthwhile area of study, and following a few weeks of intense study, they would have gained a sufficiently advanced grasp of the subject to come up with at least one entirely new theorem.

But much of their work was impenetrable, because they had their own language that they used to communicate between themselves. Very few outsiders managed to break in, apart from one or two PhD students such as Ali's partner Patrice, although her thesis turned out to be one of the few pieces of mathematics that was actually dangerous, and it was generally accepted that when I accidentally destroyed the only remaining copy by hurling it into the blades of a helicopter, I had actually done mankind a service.

One of the few other people to form a working relationship with the Vavasors was the financier Rufus Fairbanks, who persuaded them to develop some algorithms for him to help launder money on behalf of his business associates, the Gretzkys. This went surprisingly well, in that no one ever suspected that something dodgy was going on, at least not until Archimedes killed first his brother and then himself in a fit of jealous rage when he found out that he was not the only Vavasor having an affair with Fairbanks's wife.

Following their death, their papers disappeared into the safe custody of their younger brother Isaac for the next ten years, with only a select few being afforded the occasional glimpse of the contents. This only succeeded in pouring ever more fuel onto the blaze of speculation as to what was behind the double killing and what was actually in the papers.

Eventually, poor George Burgess was entrusted with the papers in order to write the definitive biography that would solve the mystery of the Vavasors, or at least divert people's attention from the real solution. And then George left his briefcase on a train on the night he was murdered by Rufus Fairbanks, which is how the papers came to be in my possession for the first time.

By now, the Gretzkys' trading strategies were fast becoming outdated. Alexei was a bright kid and a promising mathematician and one day, perhaps, he might be in a position to come up with something that came close to the quality of the Vavasors'. If he just happened to be in possession of the Vavasor papers, however, that would be a different matter, so when his father heard on the grapevine that the papers had fallen into my, and hence Dorothy's, possession, that seemed to present an opportunity. Thus Bruno Gretzky himself came to England to head the operation to steal them, and that's how Dorothy ended up being kidnapped.

Meanwhile, Julian Gowers and the Institute for Progress and Development were looking for a way to implement their global chaos agenda and were impressed with the work that Alexei was doing and proposed a partnership. However, Bruno Gretzky wasn't playing ball, so they decided to take the opportunity of his unexpected arrival in the UK to remove him. In the middle of all this, I ended up delivering the papers to Sergei and the Institute's bomb to the Gretzkys.

With Bruno out of the way, the Institute tied up their deal with Alexei, and Sergei went into hiding, only for the Petrovs to steal the Vavasor papers from him so that they could implement their own similar strategies. But they needed a mathematician to interpret the papers, which is how Dr Rory Milford, in all his various incarnations, came to be involved.

And that, of course, is how I ended up with the papers in my possession for the second time. And now I had destroyed them.

Lunch was a subdued affair. It hadn't helped that just before we sat down to eat the medics arrived to take Julian Gowers's body away, leaving a large red stain in the courtyard that the chickens took much more interest in than was seemly. But the destruction of the Vavasor papers hung over us like a pall.

'Maybe the world is better off without them,' I ventured.

'Shut up, Tom,' said Dorothy.

'I mean, that's what Patrice said about her thesis and that was something she'd spent four years on herself.'

'I said, shut up, Tom.'

'Righto.'

I glanced across at Mikhail and Arkady in turn, and they both shook their heads back at me. I wasn't going to get any help from them either. After lunch, it turned out there was an evening flight back to London that we could just get if we got a move on, and I managed to bag a ticket. Mikhail drove us to the airport in silence.

When we arrived at Minsk International 2, Dorothy grabbed her bag and stormed off into the terminal. I got out of the Suzuki and Mikhail got out as well. I turned to him and shook his hand.

'Don't worry, Tom,' said Mikhail. 'Mostly you did well. The rest is… unfortunate.'

I tried to think of some kind of witty response, but there wasn't one to be had. I just smiled and walked off with a mumbled 'thanks'. Dorothy was ahead of me in the check-in queue and I decided to wait a few minutes so that we didn't end up sitting together on the flight. Nothing I said was going to help, so it would be best if neither of us was in a position to say anything to each other for the time being.

When the plane touched down at Heathrow, I had no baggage to wait for, so I went to one of the Foreign Exchange booths and offloaded some of my roubles in order to pay for the tube. I had no idea whether Dorothy had come by car or public transport, but either way I felt it best to leave her to it. Either way, I got back to the flat ahead of her. I took out my keys and was surprised to find that she hadn't changed the locks. For an instant I felt optimistic again about our relationship, and then I remembered that I hadn't factored in the effect of my latest transgression. I was a dead man walking.

I was tired and it was late, so I found a blanket and made a bed for myself on the sofa in the living room. I was asleep in seconds and didn't even hear Dorothy come home.

'Are you sure you're OK doing this?' I said.

'Stop asking, Tom,' said Dorothy. 'Let's just get this over at least.'

Two days after we got back from Minsk, I'd succeeded in re-establishing speaking terms sufficiently to be able to discuss the possibility of a visit to the Margate safe house. I felt that this might remove one obstacle to the resumption of our relationship, by proving to Dorothy that Matheson had indeed held me against my will. Having established that, there was a decent chance that Dorothy would actually believe that Matheson had forged all those WhatsApp messages.

'Shouldn't take us too long to get there anyway,' I said. 'Provided the Dartford river crossing is open and the traffic on the M2 isn't too bad.'

The Dartford river crossing was closed because of high winds, so everything was diverted to one of the tunnels, and there were roadworks on the M2. The journey took us over three hours in the end and we were both in a foul mood by the time we got to the outskirts of Margate.

'OK, where are we going now?' said Dorothy.

'The Bubblebrook Farm Estate. Spring Bank Road.'

'Postcode, Tom. Please.'

'Oh, right.' I gave Dorothy the postcode and she tapped it into the car's navigator. Ten minutes later, we entered the Bubblebrook Farm estate, an array of identical red-brick semi-detached houses built in the eighties and already showing signs of premature ageing. The monotony was only broken up by the occasional aspirational cul-de-sac of four-bedroom detached properties and a parade of shops, two of which were boarded up.

Finally, we found the turning for Spring Bank Road and turned in, parking outside number 64.

'Well,' I said. 'This must be it.'

'Doesn't look like a safe house,' said Dorothy.

'You're not supposed to be able to tell. That's the point.'

'Hmmm.'

'No, this is definitely it. I recognise that tree over there.' I pointed to a spindly weeping willow in the front garden of the house opposite.

'So where's your room?' she said.

'It would be that one up there,' I said.

'Where are the bars? I thought you said there were bars on the windows.'

'Maybe they've taken them down.'

'Maybe they weren't there in the first place.'

'Oh, come on. Why would I lie to you?'

'I could think of a few reasons.'

'Jesus.'

This was absurd. I walked up to the front door and rang the bell. There was no response, so I rapped hard on the knocker as well. By now, Dorothy had joined me and was watching me with her arms folded.

Eventually, someone came to the door. It was an elderly lady, probably well into her eighties.

'Can I help you?' she said.

'Yes, please,' I said. 'Look, I'm sorry to bother you, but I need to speak to Helen Matheson.'

The woman looked puzzled. 'There's no one called – what was it? – Mattison here.'

'Of course there is. She kept me here as a prisoner. A month or so ago. I need to speak to her now.'

The woman seemed completely baffled.

'I have absolutely no idea what you're talking about. There's no lady called Mattison—'

'Matheson—'

'—here.'

'Oh, hang on,' I said. 'Maybe you've just moved in. That must be it.'

'Young man, I've lived here since it was built. What a strange idea.'

'Tom,' said Dorothy. 'Come on, Tom.'

'No,' I said. 'This woman's obviously got some problem with remembering things.'

'I beg your pardon?' she said. 'Are you trying to suggest I'm going doolally?'

'Well, if you don't know about Matheson, maybe you are,' I said. This was quite ridiculous.

'Just drop it, Tom,' said Dorothy, grabbing hold of one of the folds of my jacket. 'Let's forget the whole thing and go home.'

'No,' I said, shrugging her off. 'We've come this far, so we might as well finish the job. This woman's either losing it or she's covering for Matheson.'

'Fine then,' said Dorothy. 'Have it your own way. I'm off.'

'Hey wait—' I began, but she'd already started to march down the path towards the car. I turned back towards the door, which was now closing. I jammed my foot in it. 'Look,' I said. 'Can I just come in and look around?'

'No, you most certainly cannot,' said the woman.

The hell with it. I was being taken for a mug here. And then I had an inspiration. I proceeded to sniff the air in an excessively theatrical manner.

'Can you smell something?' I said, wrinkling my nose.

'I'm sorry?'

'I can definitely smell something. I think it's gas.'

'Are you sure?' There was a slight uncertainty in her question. I was in.

'I mean, I don't want to be funny,' I said, 'but you probably won't pick it up.'

'I don't think—'

'Look, do you mind if I just check? Better safe than sorry and all that.' I began to move into the house and she stepped aside to let me go through. I sniffed the air again.

'I think it's coming from up there,' I said.

'But—'

'Yes, it's definitely up there.' Before she could stop me, I jogged up the stairs and paused at the top, trying to orient myself. I closed my eyes and tried to imagine my blindfold journey out of my room. Yes, this was definitely right.

This is where Brett used to sit. I could picture him now. And there was the door to my room. I opened it and went in. The flowery wallpaper was exactly as I remembered it, as was the bed linen and the net curtains. I went over to the window, opened it and looked out to see Dorothy getting into her car. I waved frantically at her and called out, 'This is definitely it!' to her, but it was too late. She'd driven off without a backward glance.

I turned around to see the woman standing in the doorway, with a stern expression and brandishing a rolling pin.

'This is definitely the place,' I said to her. 'This is the room I was held in.'

'Please get out of my house before I call the police,' she said.

I was about to say something else when I finally realised what this must look like. I suddenly felt ridiculous.

'Sorry,' I said, sagging slightly. 'I've had a difficult few weeks.'

She didn't say anything.

'I really thought... maybe it's a different house... but the wallpaper... I mean, are you sure?'

The woman continued to look at me without saying a word.

'OK,' I said. 'I guess that's it, then.' I walked towards the door and the woman stepped aside to let me pass. I went slowly down the stairs and opened the front door. I turned round and looked up to see the woman at the top of the stairs, still with the rolling pin in her hand.

'Really sorry,' I said again. 'I don't know what's going on at all.'

It took me half an hour to find my way to the seafront. I bought myself a portion of fish and chips and found somewhere to sit overlooking the beach. I'd got halfway through when a passing gull took an interest in my lunch and decided to challenge me for it. I stood up and tried flapping my hands at the interloper, but it was armed with a beak and all I had to defend my territory was a plastic fork.

'Have it, mate,' I said, eventually, throwing the styrofoam container down on the ground. 'It's all yours.' The bird gave me a look of triumph and began tucking into the rest of my food. After a while, some of its compatriots joined in and I gained some pleasure from the fact that an unholy brawl was very soon breaking out between them all.

I sat down again and took out my phone. Then I fished in my pocket for the scrap of paper with the number on it. I hesitated for a moment and then called it.

'Hello?' said the voice.

I didn't say anything.

'Oh, it's you,' said Matheson. 'I wondered if you'd call.'

'What's going on?' I said.

'You know that would take far too long to explain, Tom.'

'You've ruined my life.'

'Oh, sweetheart,' she said. 'Do you really think I have? I think you're perfectly capable of ruining it for yourself.'

That seemed unfair. 'Who are you working for?' I said.

'You know I can't tell you that.'

'Can't you even give me a hint of what's going on?'

'Gosh, you do sound desperate.'

'I am, Matheson.'

'Helen, please.'

'Just tell me.'

'There are games being played that don't concern you, Tom.'

'They may be games, but they affect everyone else's lives. Not just mine, either.'

'Doesn't stop them being games to the players.'

'So who's winning?'

'It's not about the winning, Tom. It's about the playing.'

God, this was exasperating. 'How do you sleep at night?' I said.

'Very well, thank you. Now, is there anything else you want to say to me? Any insults?'

'I don't think so. You're despicable.'

'That sounded like an insult, darling. I feel so wounded.'

'I doubt if you've ever been wounded in your life.'

'You have no idea, Tom.'

I ended the call and blocked the number. From now on, I was going to have nothing more to do with her. It was time to get back into normal life again. Dorothy and I would almost certainly go our separate ways and I would try to find myself an ordinary job in public relations if anyone would have me. I would not be proud. I would take a lowly job if a lowly job was all that was available. I would knuckle down and toe the corporate line. I would go out for drinks with the lads and laugh along with all the awful jokes and find myself a girlfriend who did normal girlfriend stuff. I would be sensible with my money and save up for a deposit on a property that I would never be able to afford. And I would say nothing about the time I once jumped out of a burning building into the Belarusian night.

I felt something on the top of my head. I reached up to wipe it away with my hand and realised too late what it was. Up above, a gull was flapping its way towards the sea, laughing at me.

Chapter 25

'So what do you reckon?' I said.

Dolores looked at me, stretched her neck to its fullest extent and then cocked her head on one side as if to say, 'Don't ask me, mate, I'm an alpaca.' Her companion Steven had already decided not to take part in the conversation and was busy grazing.

I hadn't really expected much by way of an answer, to be honest, but there was always a chance that the act of vocalising my thoughts out loud might help. However, I'd been looking in on the alpacas every morning for the best part of two weeks now and the question still hadn't really moved on. It had become a pleasant ritual in itself and maybe that was the point. After my wild ride through Belarus, it was possible that a few weeks of unfocused pottering about was precisely what I needed.

My father still hadn't contacted Margot Evercreech about returning Dolores and Steven to her, which is why they were still here in Mad Dog McFish's field. In actual fact, the field didn't really belong to Mad Dog McFish, although the details of the arrangement that he had with one of the local farmers were somewhat opaque, involving some kind of consultancy relating to one of the farmer's special crops that he kept hidden well away from the local constabulary.

Mad Dog was something of an enigma all round. No one seemed to know what his real name was or indeed how he'd come to be known as Mad Dog, given that he was a free-loving vegan who was stoned out of his head most of the time, although there was a persistent rumour going round the caravan park that he'd once killed a man. However, the alleged homicide in question was reckoned by most to be entirely justified, given that the victim had been consistently coming in a bar too early with the tambourine in a performance of 'Give Peace a Chance'.

'No ideas, then?' I said. Dolores didn't bother responding to this at all and turned away to join Steven. She wasn't that great a conversationalist herself this morning, or maybe she was just missing Margot. I wondered if I should give my father another nudge to give her a ring. I would have called her myself if it weren't for the fact that if I spoke to her right now I might be tempted to apply for Benjamin Unsworth's old job for want of something to do.

'Bye then,' I said to no one in particular, and turned to go.

There are few certainties in life, but one thing you can rely on is that whatever you've done, however badly you've screwed up, a cat will always treat you exactly the same. Depending on the present contents of the cat's belly, that treatment will be either the purest expression of unconditional love and affection that you will ever know or sheer, undistilled contempt. When I reached my father's static caravan, I was very much on the receiving end of the latter treatment, so I took it that breakfast had already been served.

'Hello, μ,' I said, bending down to scratch the little cat's neck. She gave me a look that informed me that I shouldn't expect to get around her that easily and anyway why was I still lingering around here? Didn't I have a home to go to?

'Fair enough,' I said.

The door opened and Wally the dog bounded out, anxious to get the morning's first noseful of my groin. He was a much less complicated proposition than μ, but I felt he had work to do on his perceptions of other peoples' personal space.

'Morning, son,' said my father. 'Everything all right with the llamas?'

'Alpacas, Dad,' I said.

'Whatever. What do you want for breakfast? We're a bit low on the beluga caviar, but there's plenty of cereal in the cupboard.'

'It's OK, I'll sort myself out.'

I found a chipped bowl and poured myself some cornflakes. My father wasn't eating, as his way of getting the day going was to roll himself a hideously strong cigarette and kickstart his lungs with a bout of supercharged coughing. I'd tried to convince him that this wasn't going to end well, but the habits of a lifetime were going to take more than an intervention from me to change.

We sat for a while in companionable silence.

'Mmmm,' said my father eventually. 'Be nice to have a bit of music playing now.'

'Look,' I said. 'I'm sorry I forgot about the cassette player.'

'What cassette player?'

'You asked me for one. To replace the one we broke. And I forgot.'

'Well, would've been nice. But there you go. These things happen.'

'I've let a few people down recently,' I said.

'Son, if I ever were to sit down and write a list of all the people I've ever let down, I wouldn't be able to afford enough paper.'

Well, that was certainly true. My father's greatest redeeming feature was that however badly you'd cocked things up, you just knew he'd done far worse in his time. There was another long silence between us.

'You're still upset about her, aren't you?' he said, eventually. 'That what's-her-name. Dorothy.'

I sighed. 'I guess I am,' I said. My father still didn't know the full details of what had happened between us and probably never would. I'd tried relating the story a few times, before realising it was a futile task and giving up.

'Y'know, I often wonder about getting back together with your mother,' he said, leaning back and staring at the roof of the caravan.

'Oh, come on,' I said, before I could stop myself. 'First of all, she's living on the other side of the world, and secondly, she's sensible. I mean, no offence, Dad.'

'None taken, son.'

There was an awkward silence.

'Look,' I said. 'I didn't mean to be abrupt, but—'

'Like I said, son. No offence taken.' He took a long drag of his cigarette and followed this with a truly stupendous coughing fit that at one point seemed quite likely to bring up an entire lung. I gave him a concerned look, but he waved me away as if this was an everyday occurrence.

My father regained his composure and looked me firmly in the eye. 'You'll find a way to get her back,' he said.

'I wish I could believe you,' I said. 'She's everything I didn't know I wanted. But then again, there's a voice in the back of my head that keeps asking me if what I really need is someone a bit more normal.'

'There are better things to be in life than normal, son,' said my father. In my heart of hearts, I desperately wanted to believe this too, although I would have preferred it to have come from a more trustworthy source. If there was anyone who could have actually benefitted from an injection of normality, it was my father.

'Women, eh?' I said.

'Can't live with them, can't live without them.'

It was comforting to be exchanging tired platitudes about the battle of the sexes with someone who had not only lost even more heavily than I had but had also been disqualified on more than one occasion. But was I really any better? I had convinced myself that at least I meant well, but didn't everyone? After all, even Anya probably thought she was doing the right thing.

A slight furry pressure on my knee and a new, all-pervading canine odour alerted me that Wally had joined the conversation and was sitting at my feet underneath the table. He was looking up at me as if to say, 'I believe in you, Tom.'

'I know you do,' I thought. 'But you're a dog. It's far more complicated for us humans. Sometimes, things just seem impossible.'

Wally continued to gaze at me in adoration, albeit tinged with the slightest hint of disappointment at the extent of my self-doubt. Who knows, maybe for once he and my father were right. Perhaps I really could win Dorothy back. Somewhere, sometime, somehow, I would do something that would convince her once and for all that I really was the one for her. I just had to believe in myself and grasp that opportunity when it eventually presented itself. And given the course that my life had taken lately, there was every chance that one day it would.

Maybe.

THE END

Acknowledgements

Thanks as ever to my wonderful family Gail, Mark and Rachel for their support and encouragement. Many thanks also to Eugenia Rebotunova and Irina Nesvetova for checking over my attempts at Belarusian – any errors remaining are entirely my responsibility. Finally, massive thanks as always to my brilliant editor Abbie Headon and the rest of the team at Farrago for everything they did to make this book the best it could possibly be.

Also Available

The Truth About Archie and Pye

A Mathematical Mystery, Volume One

Something doesn't add up about Archie and Pye...

After a disastrous day at work, disillusioned junior PR executive Tom Winscombe finds himself sharing a train carriage and a dodgy Merlot with George Burgess, biographer of the Vavasor twins, mathematicians Archimedes and Pythagoras, who both died in curious circumstances a decade ago.

Burgess himself will die tonight in an equally odd manner, leaving Tom with a locked case and a lot of unanswered questions.

Join Tom and a cast of disreputable and downright dangerous characters in this witty thriller set in a murky world of murder, mystery and complex equations, involving internet conspiracy theorists, hedge fund managers, the Belarusian mafia and a cat called μ.

Also Available

A Question of Trust

A Mathematical Mystery, Volume Two

Life is not going smoothly for Tom Winscombe.

His girlfriend Dorothy has vanished, taking with her all the equipment and money of the company she ran with her friend Ali. Now Tom and Ali are forced into an awkward shared bedsit existence while they try to work out what she is up to.

Tom and Ali's investigations lead them in a host of unexpected and frankly dangerous directions, involving a pet python, an offshore stag do and an improbable application of the Fibonacci sequence. But at the end of it all, will they find Dorothy – and will she ever be able to explain what exactly is going on?

Also Available

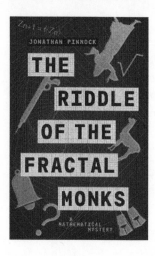

The Riddle of the Fractal Monks

A Mathematical Mystery, Volume Three

A mystery lands – literally – at Tom Winscombe's feet, and another riotous mathematical adventure begins…

Tom Winscombe and Dorothy Chan haven't managed to go on a date for some time, so it's a shame that their outing to a Promenade Concert is cut short when a mysterious cowled figure plummets from the gallery to the floor of the arena close to where they are standing. But when they find out who he was, all thoughts of romance fly out of the window.

Just who are the Fractal Monks, and what does Isaac, last of the Vavasors and custodian of the papers of famed dead mathematical geniuses Archie and Pye, want with them? How will other figures from the past also demand a slice of the action? And what other mysteries are there lurking at the bottom of the sea and at the top of mountains? The answers lie in *The Riddle of the Fractal Monks*.

About the Author

Jonathan Pinnock is the author of the novel *Mrs Darcy Versus the Aliens* (Proxima, 2011), the short story collections *Dot Dash* (Salt, 2012) and *Dip Flash* (Cultured Llama, 2018), the bio-historico-musicological-memoir thing *Take It Cool* (Two Ravens Press, 2014) and the poetry collection *Love and Loss and Other Important Stuff* (Silhouette Press, 2017). *The Truth about Archie and Pye*, the first novel in his Mathematical Mystery series, was published by Farrago in 2018. Jonathan was born in Bedford and studied Mathematics at Clare College, Cambridge, before going on to pursue a moderately successful career in software development. He also has an MA in Creative Writing from Bath Spa University. He is married with two slightly grown-up children and now lives in Somerset, where he should have moved to a long time ago.

Note from the Publisher

To receive updates on new releases in the Mathematical Mystery series – plus special offers and news of other humorous fiction series to make you smile – sign up now to the Farrago mailing list at farragobooks.com/sign-up.